Wireless Powered Communication Networks

Abbas Jamalipour • Ying Bi

Wireless Powered Communication Networks

From Security Challenges to IoT Applications

 Springer

Abbas Jamalipour
The University of Sydney
Sydney, NSW, Australia

Ying Bi
The University of Sydney
Sydney, NSW, Australia

ISBN 978-3-030-07463-0   ISBN 978-3-319-98174-1   (eBook)
https://doi.org/10.1007/978-3-319-98174-1

This Springer imprint is published by the registered company Springer Nature Switzerland AG
The registered company address is: Gewerbestrasse 11, 6330 Cham, Switzerland

Preface

This book introduces wireless powered communication networks (WPCNs) as a promising paradigm to overcome the energy bottleneck suffered by traditional wireless communication networks as well as emerging Internet-of-Things (IoT) networks. It selectively spans a coherent spectrum of fundamental aspects in WPCNs, such as wireless energy transfer (WEH) techniques, radio frequency (RF) energy harvesting receiver model, simultaneous wireless information and power transfer (SWIPT) as well as the rate-energy tradeoff arising from the joint transmission of information and energy using the same waveform. It also reviews existing literature on the network models for WPCNs, including the baseline and dual-hop WPCN models and a variety of related extensions. This book further examines the key factors including throughput, fairness, and security that must be taken into account for impeccable operation of WPCNs. The new IoT applications are targeted as a key element in those factors.

The steady growth in wireless communications, fostered by the prosperity of mobile services, has resulted in an unprecedented awareness of the potential of physical layer security (PLS) to significantly strengthen the security level of current systems. The fundamental idea of PLS is to exploit the inherent randomness of fading and interference of wireless channels to restrain the amount of information that can be gleaned at the bit level by a passive eavesdropper. Although having been studied extensively in various kinds of networks such as wireless sensor networks, cellular networks, cognitive networks, it is fair to acknowledge that the practicality of PLS in WPCNs is still an open problem.

This is the first book examining the current research to provide a unified view of wireless power transfer (WPT) and information transmission in WPCNs from a PLS perspective. Focused on designing efficient secure transmission schemes, analyzing energy evolution process, and evaluating secrecy outage performance under different channel state information (CSI), the results presented in this book shed light on how to best balance security and throughput with prudent use of harvested energy in WPCNs.

From the IoT point of view, this book highlights the opportunities and challenges for the lately emerged WPCN to seamlessly integrate into the emerging IoT ecosystem. It specifically addresses the maximization problem of uplink and downlink sum-throughout in a dual-hop WPCN, while taking fairness among WPCN users as a constraint. The results provided in this book reveal valuable insights into improving the design and deployment of future WPCNs in the upcoming IoT environment.

This book could not be completed without contributions from our graduate students at the Wireless Networking Lab, Center of Excellence in Telecommunications, the University of Sydney. Our Lab's former graduate student, Parisa Ramezani, has contributed to some of simulations and analysis provided in Chap. 3. We would like to acknowledge all members of the Lab and the Center of Excellence. We would like also to thank our respective families for their patience while we were preparing the materials and scripts for this book. The book could not be materialized without the precious support from Springer International Publishing AG and in particular Susan Lagerstrom-Fife, as well as the production editors at Springer Nature. Finally, we thank all researchers who will read this book, and we hope that the book provides a useful text in their future research endeavors.

Sydney, Australia   Abbas Jamalipour
July 2018   Ying Bi

Contents

Acronyms

AF	Amplify-and-forward
AMI	Advanced metering infrastructure
AMR	Advanced meter reading
AN	Artificial noise
AnJ	Accumulate-and-jam
AP	Access point
ATT	Accumulate-then-transmit
AWGN	Additive white Gaussian noise
c.d.f.	Cumulative distribution function
CJ	Channel state information
CR	Cooperative jamming
CSI	Cooperative relaying
DF	Decode-and-forward
DEH	Dedicated energy harvesting
DoS	Denial-of-service
ER	Energy receiver
ESS	Evolutionarily stable strategy
FD	Full-duplex
HD	Half-duplex
HES	Hybrid energy storage
i.i.d.	Independent and identically distributed
IR	Information receiver
IT	Information transmission
JCRnJ	Joint cooperative relaying and jamming
LOS	Line-of-sight
MC	Markov chain
MIMO	Multiple-input multiple-output
MRC	Maximum ratio transmission
NAN	Neighborhood area network
OEH	Opportunistic energy harvesting
PB	Power beacon

p.d.f.	Probability distribution function
PS	Power splitting
QoS	Quality-of-service
RF	Radio frequency
RTIS	Real-time incentive scheme
RTP	Real-time pricing
RV	Random variable
SIMO	Single-input multiple-output
SIMOME	Single-input multiple-output multiple-eavesdropper
SINR	Signal-to-interference-plus-noise ratio
SNR	Signal-to-noise ratio
SOP	Secrecy outage probability
SWIPT	Simultaneous wireless information and power transfer
TDMA	Time division multiple accessing
TS	Time splitting
WAN	Wide area network
WPCN	Wireless powered communication network
WPT	Wireless power transfer
ZF	Zero-forcing

Chapter 1
Introduction to Wireless Powered Communication Network

Abstract Wireless power transfer (WPT) plays a critical role in relaxing concerns related to limited operational lifetime of wireless networks. Different from traditional network devices, which rely on batteries for their energy need, devices in wireless powered communication networks (WPCNs) are able to scavenge energy from radio-frequency (RF) signals. As such, it eliminates the burden of battery recharging and/or replacement and hence provides networks with theoretically perpetual lifespans. However, due to the dramatic growth of wireless data traffic and the rapid movement towards the so-called Internet of Things (IoT), WPCNs are facing security and throughput challenges in which the traditional mechanisms are not sufficient to satisfy the user requirements. Its network performance is therefore compromised. In this chapter, we first provide an overview of the WPCNs by introducing the background of WPT, followed by a summary of the research conducted in the field. We then describe the physical-layer security (PLS) problem in WPCNs, including the causes and the impacts of the problem on the performance of WPCNs. At last, we close this chapter by discussing the applications of WPCNs in the IoT.

1.1 Overview of Wireless Powered Communication Networks

1.1.1 Wireless Power Transfer

The initial work on wireless power transfer (WPT) dates back to more than a century ago [1, 2], when Nikola Tesla used the microwave technology to do experiments on WPT and demonstrated the transmission of wireless power over a distance of 48 km. He then lit up a bank of 200 light bulbs and ran one electric motor by transmitting 100 million volts of electric power wirelessly over a distance of 26 miles. He also achieved another breakthrough by inventing the "Tesla coil" which produces high-frequency and high-voltage alternating currents and constructing the "Tesla tower" as a wireless transmission station for electrical energy transfer through the Ionosphere. After Tesla, the contributions in the WPT field were limited until William C. Brown successfully converted microwave energy to DC power via

© Springer Nature Switzerland AG 2019
A. Jamalipour, Y. Bi, *Wireless Powered Communication Networks*,
https://doi.org/10.1007/978-3-319-98174-1_1

a rectenna in 1960s and powered a model helicopter completely through microwave power transfer [3]. In 1975, he beamed 30 kW microwave power over a distance of 1 mile, achieving an efficiency of 84% [4].

Despite the early efforts on wireless power transmission, serious steps toward the widespread realization of this technique were not taken until recently when the rapid advances of electronic devices and the need for on-demand and cable-free energy transmission motivated the research community and the industry to pay earnest attention to the development and commercialization of WPT techniques. Generally, WPT can be in the form of inductive coupling [5], magnetic resonant coupling [6], or RF-enabled WET. The former two are near-field wireless charging technologies, where the generated electromagnetic field dominates the region close to the transmitter or scattering object and the power is attenuated according to the cube of the reciprocal of the charging distance [7]. In contrast, RF-enabled WPT is a far-field charging technology used for transferring wireless energy over long ranges, in which the power decreases with the square of the reciprocal of the charging distance.

Inductive coupling refers to the energy transfer from one coil to another as a result of the mutual inductance between the two coils. It occurs when an alternating current in the transmitter coil generates a magnetic field across the terminals of the receiver coil. This magnetic field induces voltage in the receiver coil which can be used for powering devices. This technique is very efficient when the magnetic coupling between the two coils is large enough, i.e., when the transmitter and the receiver coils are close to each other. Magnetic resonant coupling uses the principle of resonance for increasing the energy transfer range and efficiency. Indeed, if the two coils are tuned at the same resonant frequency, they can exchange energy with greater efficiency at a longer operating distance compared to inductive coupling [8]. However, charging distance with magnetic resonant coupling is still limited to a few meters which makes it inapplicable for mobile and remote charging.

In RF energy transfer, radio signals with frequency range from 3 kHz to 300 GHz are used as a medium to carry energy in the form of electromagnetic radiation [9]. Compared to inductive coupling and magnetic resonant coupling, RF-enabled WPT can operate in longer ranges thanks to radiative properties of electromagnetic waves. The harvested RF power in the free space can be calculated using the Friis equation [10] as follows:

$$P_R = P_T \frac{G_T G_R \lambda^2}{(4\pi d)^2}, \qquad (1.1.1)$$

where P_R is the received power, P_T is the transmitted power, G_T is the transmit antenna gain, G_R is the receive antenna gain, λ is the wavelength used, and d is the distance between the transmit antenna and the receive antenna.

Figure 1.1 shows the block diagram of an RF energy harvesting system. The antenna collects RF signals from available RF sources. The matching circuit is used to ensure that the maximum RF power is delivered to the rectifier, which then converts RF power to DC power. And finally, the DC power is stored in the

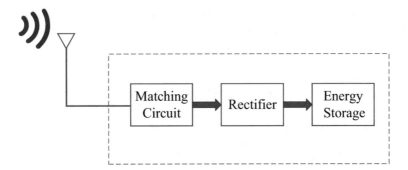

Fig. 1.1 Block diagram of an RF energy harvesting system

energy storage which can be either a rechargeable battery or a super-capacitor. The efficiency of the RF energy harvester depends on the efficiency of the antenna, the accuracy of the matching circuit between the antenna and the rectifier, and the efficiency of the rectifier that converts the received RF signals to DC voltage [11].

RF energy harvesting sources can be classified into two categories [12]: ambient RF sources and dedicated RF sources. Ambient RF sources are not intended for RF energy transfer and are freely available in the environment [13, 14]. TV and radio towers, WiFi access points, mobile phones, and mobile base stations are some of the ambient RF energy sources available around us. As another example, in a cognitive radio network, secondary users can take advantage of the primary users' transmissions as an ambient RF source to harvest their required energy [15, 16]. On the other hand, dedicated RF sources are specifically intended for on-demand RF energy transfer and are more suitable for applications with quality of service (QoS) constraints due to their high power density and controllable behavior. The TX91501 Powercaster transmitter is an example of a dedicated RF energy source which broadcasts radio waves in the unlicensed 915 MHz ISM band [17].

With this brief background on WPT techniques, we are now ready to review two prominent topics of research in this area which constitute the basis of our work.

1.1.2 Simultaneous Wireless Information and Power Transfer

Simultaneous wireless information and power transfer (SWIPT) is a spectrum-efficient method for powering energy-constrained devices. Instead of occupying the spectrum for separate energy and information transfer, SWIPT technique utilizes the same RF signal for both energy harvesting (EH) and information decoding (ID). This objective is achievable because any information-bearing signal also carries energy that can be harvested by either the information receiver who exploits the received RF signal for both EH and ID or ambient devices who use the RF signal only for collecting their needed energy. However, a practical limitation exists for implementing a SWIPT receiver who wants to utilize the same RF signal for both purposes because the energy harvesting operation performed in the RF domain destroys the information content [18]. For this reason, some practical SWIPT receiver architectures have been proposed in the literature which will be reviewed later in this section.

The idea of transmitting energy and information simultaneously was first introduced in [19] for noisy single-input single-output (SISO) channels. The author proposed a capacity-energy function to characterize the trade-off between energy and information transmission rates. This work has been extended to frequency-selective channels and multiple-access channels in [20] and [21], respectively. These early studies have assumed an ideal SWIPT receiver implementation in which both information and energy can be extracted from the same radio frequency signal at the same time.

Reference [22] addressed the practical constraint of a SWIPT receiver for simultaneous EH and ID from the same RF signal. Indeed, any information embedded in the received signals sent to the EH circuit is lost during the EH process. Accordingly, two practical receiver implementations have been proposed in [22], namely, time switching (TS) receiver and power splitting (PS) receiver depicted in Fig. 1.2a and b, respectively.

In a TS receiver, time is divided into two orthogonal time-slots. The receiver periodically switches between EH and ID. In this receiver setup, the transmitter can optimize the waveforms for EH and ID in the corresponding time-slots because there is a fundamental difference in the optimal waveforms for information and energy transmissions [23]. A PS receiver differs from a TS one in that both EH and ID are performed at the same time. To achieve this, the received signal power is split into two streams with power ratios ρ and $1 - \rho$ to be fed into the energy harvester and the information decoder, respectively. Different rate-energy trade-offs can be obtained by changing the time-slot durations in the TS architecture and the power splitting ratio in the PS receiver.

In [24], the SWIPT receiver setups proposed in [23] have been generalized to a dynamic power splitting design in which the received signal is dynamically split into two streams with adjustable power ratio for energy harvesting and information decoding. The authors also proposed two receiver architectures, namely, separated and integrated information and energy receivers. A separated receiver splits the received signal in the RF band and transfers the two streams to the conventional energy receiver and information receiver for harvesting energy and decoding information, respectively. In the integrated architecture, the RF-to-baseband converter is replaced by a passive rectifier. Here, the signal is split after being converted to the DC current. The authors characterized rate-energy (R-E) trade-offs for both receivers taking into account the circuit power consumption.

Xiang and Tao considered a three node MISO system in [25] with one multi-antenna transmitter, one single-antenna energy receiver, and one single-antenna information receiver. Assuming that the transmitter only has imperfect knowledge of the channels, they studied the robust beamforming problem for maximizing the

Fig. 1.2 Practical receiver
designs for energy harvesting
and information decoding
[22]. (**a**) Time switching
SWIPT receiver. (**b**) Power
splitting SWIPT receiver

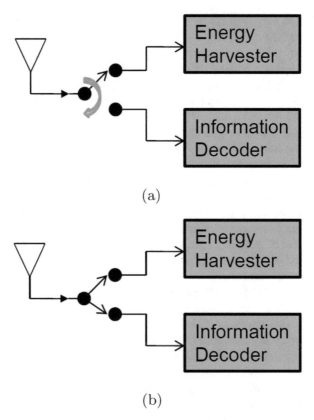

(a)

(b)

harvested energy of the energy receiver while guaranteeing a specific target rate for
the information receiver.

Reference [26] integrated SWIPT with relaying networks and proposed two
relaying protocols based on TS and PS receiver architectures. The authors considered a three-node amplify-and-forward (AF) relaying network with a source-
destination pair and a relay in between, assuming that there is no direct link between
the source and the destination. The energy-constrained relay node uses the signal
transmitted by the source node for both energy harvesting and information relaying.
The relay applies either the TS or the PS approach in order to harvest energy from
the source's RF signal and forward the contained information to the destination
using its harvested energy. The throughput performance of the proposed protocols
has been analyzed in both delay-limited and delay-tolerant transmission modes.
Their study revealed that in such a cooperative setup, locating the relay node closer
to the source results in higher throughput.

Lee *et al.* extended the work in [26] by considering the presence of a direct link
between the source and the destination [27]. The outage probability of the system

and the power-splitting factor at the relay have been found in closed form. According to their results, the cooperative scheme with direct link shows superior performance to both the non-cooperative approach (i.e., absence of the relay) and the cooperative approach without a direct link between the source and the destination.

In [28], a framework was developed for realizing SWIPT in broadband wireless systems. Utilizing OFDM, the broadband channel has been divided into orthogonal sub-channels with all sub-channels assigned to one user in a single-user system and a single sub-channel assigned to each of the users in a multi-user setup. This frequency diversity is shown to help improve the efficiency of SWIPT.

Reference [29] considered a large-scale network with multiple transmitter-receiver pairs. The author studied both non-cooperative and cooperative schemes: The former consists of a random number of energy-stable transmitters and energy-constrained receivers, whereas the latter also includes a random number of energy-stable relays which assist the energy and information transfer. The receivers employ the PS technique for collecting their needed energy. The fundamental trade-off between the outage probability and the harvested energy has been investigated in both scenarios.

Reference [30] pointed out the disadvantages of TS and PS methods and proposed an antenna switching (AS) technique for SWIPT. According to [30], the main drawback of TS is that using dedicated time slots for energy harvesting leads to a non-continuous transmission of data. What's more, TS techniques require strict synchronization as any timing inaccuracy may result in loss of information. On the other hand, PS suffers from additional complexity and cost since it requires ideal power splitting circuits. For these reasons, a low-complexity AS SWIPT protocol has been proposed for MIMO relay channels based on the principles of generalized selection combiner.

A practical scenario for SWIPT has been presented in [31], where a multi-antenna access point (AP) transmits information and energy to a single-antenna user. To model a realistic system, the authors assume imperfect channel state information (CSI) at the AP, presence of penalties in CSI acquisition, and non-zero power consumption for the receiver in CSI estimation and signal decoding procedures. Three cases have been studied: (a) no CSI at the AP, (b) imperfect CSI obtained by means of pilot estimation, and (c) imperfect CSI obtained by means of analog symbols feedback. The findings revealed that the availability of CSI knowledge at the AP is always helpful even though some resources are used in the channel estimation procedure and the resulting information is not perfect.

Reference [32] investigated SWIPT in a cognitive radio network, where a number of energy-limited SUs receive both energy and information from a secondary base station (SBS). The objective is to minimize the transmit power of the SBS by jointly optimizing the transmit beamforming vector at the SBS and the power-splitting ratios at the SUs. By doing so, the EH and QoS constraints of each SU can be met whilst the interference level to the primary network is also kept below a threshold.

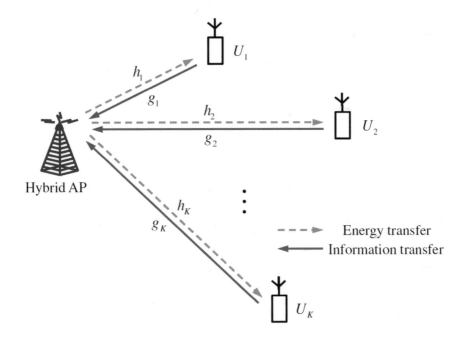

Fig. 1.3 A wireless powered communication network [33]

1.1.3   Wireless Powered Communication Networks

A wireless powered communication network (WPCN) basically consists of a hybrid access point (HAP) and a number of energy harvesting users which rely on RF energy transmission of the HAP to scavenge the needed energy for their communication. An example model is depicted in Fig. 1.3, where the HAP and the users are equipped with a single antenna each and operate in the half-duplex (HD) mode meaning that they cannot perform transmission and reception at the same time. A harvest-then-transmit protocol has been proposed in [33] which divides the frame into a wireless energy transfer (WET) and a wireless information transmission (WIT) phase as plotted in Fig. 1.4. In the first τ_0 fraction of time, the HAP broadcasts an energy signal to the users in the downlink. The users store the energy harvested during the WET phase in a rechargeable battery and then transmit their information to the HAP in the uplink by time division multiple access (TDMA) utilizing their harvested energy.

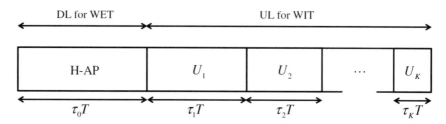

Fig. 1.4 The harvest-then-transmit protocol [33]

The time duration allocated for WET and WIT play an important part in the overall network performance. A greater τ_0 results in more harvested energy for the users leading to a higher throughput accordingly; however, as more time is dedicated for energy transfer, less time is left for information transmission which subsequently degrades the throughput. Hence, there should be an optimal τ_0 which maximizes the uplink throughput. This optimal value depends on the channel conditions between the HAP and the users. According to [33], as uplink and downlink channel power gains become larger, the optimal policy is to allot more time for WIT instead of WET. That's because when the downlink channel gets better, users can harvest sufficient energy in shorter time while in good uplink channel conditions, the users need less power to achieve the same throughput performance, both of which help reduce the energy transfer duration and save time for information transmission. Furthermore, the allocated information transmission time to each user also affects the sum-throughput. Generally, the users who can contribute more to the total throughput of the network get a larger transmission time. As a consequence, the time assigned for each user's data transmission is coupled with its uplink and downlink channel power gains which results in unbalanced time allocation among the users due to the so-called doubly near-far problem.

The doubly near-far problem stems from unequal distance of different users from the HAP. In a WPCN, users far away from the HAP receive less amount of wireless energy than near users in the WET phase, but need to transmit with greater power in the WIT phase. To achieve the best possible throughput performance for the entire network, the optimal design is to allocate a small amount of time to the far users and leave more transmission time for the users with better channel conditions. Such a strategy sacrifices the far users' throughputs for the sake of total throughput maximization and induces serious unfairness among WPCN users. To tackle the doubly near-far problem, the authors of [33] solved a common-throughput maximization problem which provides all users with equal throughput by allotting more data transmission time to further users.

The work in [33] was extended in [34] for improving energy harvesting of the users. In this work, employing a full duplex (FD) HAP has been proposed which is able to perform energy transfer and data reception simultaneously on the same frequency, allowing time and spectrum resources to be used more efficiently as compared to the HD scenario. The FD HAP transfers energy signals in all time-slots of the frame and at the same time, receives information from the users who transmit to the HAP by TDMA. In [34], each user can harvest energy in all the time-slots other than the one in which it is transmitting information to the HAP. Thus, the amount of the harvested energy of the users is substantially increased. Another pioneering work in using FD HAP is [35] which takes energy causality into account by letting users harvest energy only until their allocated information transmission slot.

The performance gain achieved by FD operation mode relies heavily on the ability of the HAP to perform self-interference cancellation (SIC). The reason is that the transmitted energy signal can severely affect the information signal of the users received by the HAP as the power of the energy signal overrides that of the attenuated information signal. The achievable throughput thus depends on the extent to which the self-interference (SI) can be canceled. Reliable SIC techniques are needed to ensure the effect of the SI is significantly reduced. So far, a number of methods for SIC have been proposed which can be classified into propagation-domain, analog-circuit-domain and digital-domain approaches [36]. A combination of the aforementioned techniques can considerably reduce the amount of SI; though at the expense of increased complexity. As reported in [34], FD WPCN outperforms the HD counterpart in terms of the overall throughput when SI can be effectively canceled.

FD operation is also possible at the user side [37]. FD users can transmit information and harvest energy at the same time which results in increased harvested energy for them. Here, there is no need for SIC and the user can treat its own transmitted signal as a viable source for energy harvesting.

When a single-antenna HAP transmits energy in an omni-directional manner, the severe signal power loss over distance leaves only a limited amount of harvestable energy for the users. For this reason, exploiting multi-antenna HAP has been proposed to improve the RF energy transfer efficiency [38–40]. When the HAP is equipped with multiple antennas, energy beamforming can be employed to design the phase and amplitude of the energy signal at each antenna in a way that the combined energy transfer performance is optimized. Not only the harvested energy of the users would be increased by this technique, but the higher efficiency of energy transfer also enables faster charging of devices [41], leaving more time for data transmission. What's more, multi-antenna HAP makes it possible for multiple users to transmit information in the uplink simultaneously using space division multiple access (SDMA) which substantially increases the throughput. [40] studied the sum-throughput maximization problem in the multi-antenna WPCN and found optimal energy beamforming, receive beamforming, and time-slot allocations. Similar to the baseline single-antenna WPCN, the sum-throughput maximization problem sacrifices some of the users' throughputs. Therefore, the total throughput of the network is maximized at the cost of low fairness arising from the doubly near-far problem. To overcome this problem and ensure fairness, references [38] and [39] optimized time and energy allocations plus beamforming vectors to maximize the minimum throughput among users under the perfect and imperfect CSI assumptions, respectively.

A large-scale WPCN has been studied in [42] in which multiple HAPs are responsible for energy/information transmission coordination to/from a large number of users. The large-scale WPCN is modeled based on homogeneous Poisson Point Processes and a scalable energy/information transfer scheme is devised to serve the large number of network users. The harvest-then-transmit protocol proposed in this work is slightly different from the one in [33] with T being the total number of time-slots for energy and information transmission and N ($1 \leqslant N \leqslant T - 1$) slots devoted to energy transfer from the HAPs to the users. Each user randomly picks a slot from the $T - N$ time-slots dedicated for information transfer and transmits its data to the nearest HAP. The objective is to find the optimal transmit power of the users plus the optimal number of energy transfer slots (N) using stochastic geometry tools in order to maximize the spatial throughput defined as sum of the throughputs of all users normalized by the network area. As reported in [42], increasing the number of users per network area results in greater maximum achievable spatial throughput; however, more HAPs need to be deployed to achieve the maximum throughput due to the increased interference level caused by the dense deployment of the users.

References [43] and [44] investigated the integration of WPCN with cognitive radio networks and presented cooperation strategies such that WPCN as the secondary network helps the primary network by relaying the primary transmitter (PT)'s data to the primary receiver (PR). Assuming that PT's message is made known to the HAP, [43] proposed to incorporate the cooperation in the WET phase. The HAP cooperatively sends PT's data to the PR using its WET signal. Higher transmit power is a bonus given to the HAP for this cooperation. Optimal time and power allocations for WET and WIT phases have been found for maximizing the WPCN sum-throughput under a primary rate constraint. Non-cooperative cognitive WPCN has also been studied in [43] in which the secondary network does not assist the primary communication, but keeps the interference level at the primary side below a certain threshold. The cooperative scheme is shown to outperform the non-cooperative approach in terms of both the primary achievable rate and the secondary sum-throughput. Reference [44] proposed another cooperation strategy, where the WPCN users are responsible for relaying PT's data. As a result of this cooperation, spectrum access is awarded to the secondary network on the condition that the target primary rate is met. Here, it is important to find the optimal set of relaying users, the amount of energy they must allocate for relaying, and the time duration dedicated to each user's data transmission. In this scheme, the secondary network throughput highly depends on the target primary rate. A greater target rate makes WPCN users spend more time and energy on relaying and therefore leaves shorter time and less energy for their own transmissions resulting in secondary throughput reduction.

In [45], Wu *et al.* studied a different WPCN setup, where instead of a HAP responsible for both energy and information coordination, ET and AP are used. This system model can eliminate the doubly near-far problem as a user which is far from the access point may be near to the energy transmitter (or vice versa). Consequently, the severe signal attenuation in uplink information transmission can be made up by better harvesting conditions in the downlink (or less harvested energy in the downlink can be made up by good uplink channel conditions). However, using separate AP and ET makes the energy/information transfer coordination more difficult and also adds to the production and operation costs [46]. Assuming an initial energy for each user and also the capability of storing the harvested energy for future use, the authors of [45] jointly optimized time allocation and power to maximize the energy efficiency of the proposed system.

Despite all the efforts made to enhance and extend the newly-emerged WPCNs, most of the works in this area are limited to single-hop communication between the HAP (or AP) and the users. References [47] and [48] are two exceptions to this single-hop model; they exploit a relaying approach by letting the information transmission of a user be assisted by either another user [47] or a relay [48]. In [47], a two-user WPCN has been considered where the nearer user to the HAP with better channel conditions dedicates a portion of its allocated time and harvested energy to relay the other user's data. Reference [48] presented a wireless powered cooperative communication network consisting of a HAP, a source, and a relay operating under the proposed harvest-then-cooperate protocol. The source and the relay harvest energy from the HAP during the downlink energy transfer phase and cooperatively transmit the source's data to the HAP in the uplink. Although these two network models can provide insights for implementing cooperative communication in WPCNs, they limit the network to a HAP and one [48] or two [47] users and do not consider a network where multiple users need to communicate with the HAP. Considering the increasing number of network users and the large dimension of IoT, WPCN with only one or two users is an over-simplified and unrealistic scenario. Hence, cooperative and multi-hop communication in WPCN need much more investigation to satisfy practical implementation needs.

Moreover, even though remarkable research has been conducted in the fields of SWIPT and WPCN during the last few years (as reviewed in the previous section and this section, respectively), integrating these two interesting research topics has not yet been considered, mainly because the downlink communication (information transmission from the HAP to the users) in WPCN has been overlooked so far. However, downlink communication is an inevitable part of wireless communication networks and the information transmission from the HAP to the users is worthy of investigation.

1.2   Physical Layer Security Challenges in WPCNs

In WPCNs, the operating power sensitivity of energy receivers (ERs) is typically much smaller than that of information receivers (IRs). Hence, only the receivers which are in close proximity to the transmitter are scheduled for RF energy harvesting, and there may be situations that ERs act as eavesdroppers to overhear the messages sent to IRs. This near-far problem gives rise to a challenging physical layer security (PLS) issue. In this section, some major challenges and progress in the literature for improving PLS in WPCNs are discussed, mainly from a signal processing perspective.

1.2.1 Fundamentals of Physical Layer Security

In a typical three-terminal wireless network, the problem of secrecy and confidentiality arises when a transmitter wants to send a secret message to an intended receiver in the presence of an eavesdropper. Based on the vast majority of research on PLS, the eavesdropper is treated as an unauthorized receiver that works in a passive eavesdropping mode; it does not transmit but only listens to conceal its presence. The wireless channel between the legitimate terminals is called the main channel, and the one between the transmitter and the eavesdropper is called the wiretap channel. When the intended receiver and the eavesdropper are not collocated, signals observed at the outputs of the main channel and the wiretap channel are usually different; this is principally caused by the physical phenomena fading and noise. In wireless systems, fading may either be caused by multi-path propagation, referred to as multi-path induced fading, or due to shadowing from obstacles affecting the wave propagation, sometimes referred to as shadow fading. Noise is the intrinsic element of almost all physical communication systems.

In an effort to investigate the role of noise in the context of secure wireless communications, Wyner introduced the wiretap channel model [49] illustrated in Fig. 1.5. This model ushers in a new era in information-theoretic security—PLS. The main characteristics of Wyner's approach are that

- without the aid of secret keys, Alice uses wiretap coding to encode messages into codewords;
- the codewords transmitted by Alice propagate over noisy channels;
- the observations of codewords at Bob and Eve are different.

Specifically, Wyner introduced two rate parameters, i.e., the codeword transmission rate R_t and the secrecy rate R_s. The secret message is encoded by generating 2^{nR_t} codewords $X^n(m; c)$ of length n, where $m = 1, 2, \ldots, 2^{nR_s}$ and $c = 1, 2, \ldots, 2^{n(R_t - R_s)}$. The positive rate difference $R_e := R_t - R_s$, also called the rate redundancy, is the coding cost of providing secrecy against eavesdropping. For each message index m, Alice randomly selects a codeword index c with uniform probability and then transmits the corresponding codeword $X^n(m; c)$ over the main channel. At the output side, Bob receives a noisy version of the transmitted

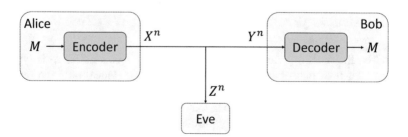

Fig. 1.5 The wiretap channel model of Wyner with noisy wireless channels

codeword, denoted by Y^n. Meanwhile, Eve observes from the wiretap channel a different but also corrupted version of the codeword, denoted by Z^n. Perfect secrecy is achievable for the wiretap channel model even in the absence of a shared secret key, provided Z^n is more corrupt than Y^n. In addition, although the use of wiretap codes make the guaranteed information rates smaller than that in cryptography, PLS places no restrictions on the computing resources of Eve, and therefore, perfect secrecy is provable.

1.2.2 Recent Advances in Improving PLS in WPCNs

The authors of [50] are among the first to investigate the beamforming schemes for secure communication in a MISO downlink system with a multi-antenna transmitter, a single IR, and multiple ERs. Besides external eavesdroppers, the ERs are also treated as potential eavesdroppers because of their decoding capability. The transmitter performs SWIPT to transmit information and wireless energy to the IR and ERs, respectively. Meanwhile, the transmitter also emits artificial noise (AN) to combat eavesdropping. With the assumed imperfect CSI of potential eavesdroppers and no CSI of passive eavesdroppers, the authors designed a resource allocation algorithm for minimizing the total transmit power radiated by the transmitter. The formulated optimization problem is non-convex and tackled by a semi-definite programming relaxation approach to obtain the optimal solution. The results provide the close-to-optimal performance of the proposed schemes and demonstrate significant transmit power savings achieved by optimizing the AN and energy signal generation. A similar worst-case based transmit power minimization problem is studied by Zhang *et al.* in [51, 52], but with a more threatening scenario where ERs collude to perform joint decoding of the confidential messages. Besides, the authors of [53] explored the transmit power minimization problem and the max-min fairness energy harvesting problem for a MISO cognitive radio network, where both the bounded CSI error model and the probabilistic CSI error model are investigated. Secrecy performance for a SIMO SWIPT system is presented in [54] in terms of secrecy outage probability and average secrecy capacity. The system model is extended to a three-node MIMO system with one IR and one eavesdropping ER in [55, 56], with the same objective of maximizing the achievable secrecy rate subject to two power constraints including a maximum transmit-side power constraint and a minimum ER-side power constraint. Motivated by cognitive radio application, the authors of [56] further extended their work by replacing the minimum ER-side power constraint with a maximum power constraint to decrease the interference at the receiver. FD techniques for secure transmission in SWIPT system are exploited in [57, 58], where the same system model is studied with a transmitter communicating with a wireless-powered FD receiver in the presence of a passive eavesdropper. A two-phase secure transmission scheme was proposed; in the first phase, energy harvesting is conducted from the transmitter to the receiver, and in the second phase, the transmitter transmits secret information under the protection of AN sent from the receiver using the harvested energy in the first phase. Closed-form expressions for the connection outage probability, the secrecy outage probability, and the transmission outage probability have been derived. Furthermore, an optimization problem has been formulated for maximizing the secrecy throughput in [57], whilst in [58] the formulated problem is to maximize the secrecy energy efficiency, defined as the total successfully and securely transmitted bits per unit time to the total energy consumed at the transmitter.

Different from the above literature, the authors of [59] coped with the problem of secure transmission in a two-hop wireless-powered relay network. In order to keep the transmitted information confidential from the untrusted relay, destination-assisted jamming has been proposed. Both power splitting (PS) and time splitting (TS) energy harvesting policies are adopted at the relay. In the proposed transmission scheme, the relay harvests energy not only from the source but also from the jamming signals sent by the destination. As such, the jamming signals are effectively utilized to protect secrecy as well as to act as an additional energy source. The simulation results illustrate that the PS policy achieves better optimal secrecy outage probability and optimal ergodic secrecy rate than that of the TS policy at higher target secrecy rate and transmit SNR, respectively. The studies in [60] and [61] consider a four-node relay network consisting of a transmitter, an intended information receiver, a multi-antenna wireless-powered relay, and an external eavesdropper. Besides the conventional TS and PS energy harvesting relay, the authors of [60] also studied the use of an ideal relay which has the capability to independently and concurrently process the information signal and to harvest energy from the same received signal. In each case, accurate analytical expressions for the ergodic secrecy capacity are derived. Also, the TS and PS factors are optimized to maximize the secrecy capacity in various system configurations. The focus of [61] is to find the optimal AN-aided secure robust beamformer for minimizing the transmission power at the relay, while guaranteeing the secrecy rate constraint. The formulated problem is solved by using the S-procedure and semidefinite relaxation techniques.

An secrecy rate maximization problem is studied in [62] for an AF relay network with multiple wireless-powered relays. The authors proposed a block-wise penalty function method to jointly optimize the PS ratio and beamformer for the active relays, whilst the rest idle relays are treated as potential eavesdroppers. The work in [63] considers the same system model as in [62] with a relay-assisted cooperative jamming (CJ) scheme. Both scenarios with global CSI and local CSI are investigated. In the global CSI case, closed-form expressions for the optimal and/or suboptimal AF-relay beamforming vectors are derived to maximize the achievable secrecy rate, subject to individual power constraints of the relays, using the technique of semidefinite relaxation. A fully distributed algorithm utilizing only local CSI at each relay is also proposed as a performance benchmark. Simulation results validate the effectiveness of the proposed multi-AF relaying with CJ over other suboptimal designs. Along this line of research, PLS in WPCNs has also been studied in other various multiuser relay channels, such as 5G network [64], massive MIMO network [65], and cognitive radio network [66].

The impact of a wireless-powered jammer on secrecy performance is first studied in [67]. The authors proposed a two-phase communication protocol: In the first phase, the source transfers energy-bearing signal to recharge the jammer. In the second phase, the source transmits the information-bearing signal to the destination; meanwhile, the jammer uses the acquired energy in the first phase to transmit jamming signals to interfere with the eavesdropper. Simulation results reveal that there exists an upper bound for the secrecy throughput when the jammer is equipped with only single antenna. In contrast, when the jammer has multiple antennas, the secrecy throughput grows unbounded with the increasing source transmit power.

Unlike [67], the authors of [68] considered a more general multi-user scenario where both the jammer and the users harvest energy from a hybrid access-point (AP) to transmit information. Specifically, the hybrid AP first transfers power to replenish the batteries of the users and the jammer in the wireless power transfer (WPT) phase. Then in the subsequent information transmission (IT) phase, each user sends information to the AP in a time division multiple access manner, whilst the jammer generates jamming signals to impair the existing multiple eavesdroppers. According to the availability of CSI of the eavesdroppers, two different secrecy metrics are used. When the perfect CSI of the eavesdroppers is known, the objective is to maximize the secrecy rate by optimizing the time allocation between the WPT phase and the IT phase. Otherwise, when the instantaneous CSI of the eavesdroppers is not available, secrecy outage probability is minimized. Shafie *et al.* extended the work in [68] by considering multiple jammers in [69]. Specifically, an ordered-based distributive jamming scheme is proposed for securing the transmission between the transmitter and an IR and powering an ER equipped with a nonlinear energy harvester. In each time slot, only one jammer out of N candidates is selected to jam the eavesdropper and power the ER. The selection criteria is that the channel gain between the selected jammer and the IR is below a threshold and that between it and the ER is above another threshold. The motivation of such selection is to make the interference at the IR as low as possible while making the harvested energy at the ER as much as possible.

There are also works that consider the transmitter rather than the jammer as an energy-constrained node. For instance, the authors of [70] studied the secure transmission from a wireless powered transmitter to a receiver, in the presence of multiple eavesdroppers. A multi-antenna node, i.e., referred to as a hybrid base station, is incorporated into the wiretap channel model to assist jamming. For such setup, the authors proposed a *harvest-then-transmit* (HTT) protocol, where each transmission block is divided into two parts: During the first τT $(0 < \tau < 1)$ amount of time, the transmitter harvests energy from the RF signals transmitted by the base station. In the remaining $(1 - \tau)T$ amount of time, the transmitter uses the harvested energy to transmit information signal to the receiver, while the base station transmits jamming signal to confound the eavesdroppers at the same time. Optimization techniques have been applied to maximize the secrecy throughput via finding the optimal time partition ratio τ. Based on the same HTT transmission scheme, the authors of [71, 72] considered a similar wireless-powered wiretap channel, where an energy-constrained source, powered by a dedicated power beacon

(PB), aims to establish confidential communication with a legitimate receiver, in the presence of one single eavesdropper. The differences lies in that [71] focuses on designing various kinds of jamming signals based on the availability of CSI, whereas [72] concentrates on investigating two popular multi-antenna transmission schemes at the source, namely, maximum ratio transmission and transmit antenna selection.

The studies that combine cooperative relaying and cooperative jamming to secure RF energy harvesting network are conducted in [73] and [74]. The authors in [73] focused on the use of multiple jammers in a multi-antenna AF relay wiretap channel. The objective is to maximize the secrecy rate subject to the transmit power constraints of the AF relay and the jammers. Based on the availability of the eavesdropper's CSI at the transmitter, the upper or lower bounds of the achievable secrecy rate are derived. In [74], FD technologies are fully exploited to secure an FD relay wireless network assisted by a wireless-powered FD jammer. By taking into account the energy consumption due to signal processing at the jammer, the authors derive a sufficient condition for the jammer to operate with reliable energy supply. Simulation results quantify the secrecy rate benefits brought by the FD relay and the FD jammer.

1.3   Applications of WPCNs in Internet of Things

With the ever-increasing speed of technological advances, more and more objects are being connected to the Internet every day, and the world is moving towards the next generation of the Internet: the so-called Internet of Things (IoT). Different definitions of the IoT exist in the literature [75]. IEEE defines IoT as a "*A network of items, each embedded with sensors, which are connected to the Internet.*" A more detailed definition of IoT is provided by Uckelmann *et al.* in the book "Architecting the Internet of Things":

> *The future Internet of Things links uniquely identifiable things to their virtual representations in the Internet containing or linking to additional information on their identity, status, location or any other business, social or privately relevant information at a financial or non-financial pay-off that exceeds the efforts of information provisioning and offers information access to non-predefined participants. The provided accurate and appropriate information may be accessed in the right quantity and condition, at the right time and place at the right price. The Internet of Things is not synonymous with ubiquitous/pervasive computing, the Internet Protocol (IP), communication technology, embedded devices, its applications, the Internet of People or the Intranet/Extranet of Things, yet it combines aspects and technologies of all of these approaches.*

Although numerous definitions of IoT exist in the literature, there is a common understanding in the research community about this concept and its high potential to change the world. IoT links the objects of the real world with the virtual world, thus enabling anytime, anyplace connectivity for anything and not only for anyone. It refers to a world where physical objects and beings, as well as virtual data and environments, all interact with each other in the same space and time [76]. An IoT object is not an ordinary object, but a smart one capable of more than just its functional specification. It can receive inputs from the world and transform those into data which subsequently are sent onto the Internet for collection and processing. It can also produce outputs into the world some of which could be triggered by data that have been collected and processed on the Internet [77]. IoT enables physical objects to see, hear, think and perform jobs by having them "talk" together, to share information and to coordinate decisions [78]. Through the use of intelligent decision-making algorithms in software applications, objects can respond appropriately to physical phenomena, based on the very latest information collected about physical entities and taking into account the patterns in the historical data, either for the same entity or for similar entities [76]. It is worth mentioning that IoT objects must be able to perform all of these tasks autonomously, i.e., without any human involvement. They should be able to self-configure, self-maintain, self-repair, make independent decision, and eventually manage their own disassembly and disposal [79]. IoT opens opportunities for the development of a wide range of application domains which include but are not limited to smart cities, e-health, logistics, and transportation. Figure 1.6 shows some of the application domains of IoT.

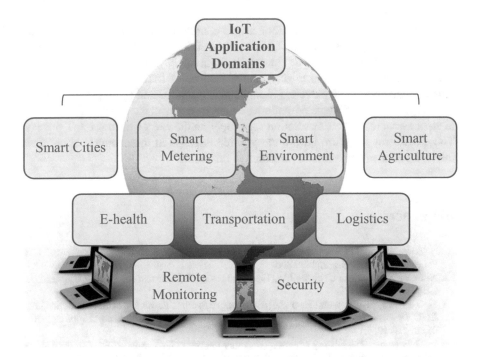

Fig. 1.6 IoT application domains

With its innumerable application scenarios, IoT is expected to improve the quality of life, create new markets, and bring huge profits for businesses. However, realizing IoT in a large scale is not an easy task and numerous challenges need to be overcome before this concept is widely accepted. Lack of standardization, privacy issues, and security concerns are among the most important challenges IoT needs to deal with.

Energy scarcity issue is another challenge that has to be surmounted for an effective and widespread adoption of the IoT paradigm. During the last two decades, extensive research has been done for designing methods and protocols for conserving as much energy as possible in order to postpone the inevitable battery depletion of wireless devices. Specifically, we have observed the trend of designing energy-efficient MAC and routing protocols for wireless sensor networks (WSNs)— key enablers of IoT—in order to keep the sensors operative for longer periods (refer to [80–85] for some well-known MAC and routing protocols of WSNs). However, the energy conservation methods proposed in the literature often trade off performance for lifetime extensions and make a compromise between quality of service and energy consumption. Even if we neglect this performance trade-off, the fact is that batteries will ultimately be depleted no matter how efficient the energy consumption is. As a consequence, battery recharging/replacement will be required which may not always be a practical solution. For example, take a health-care application scenario where sensors are implemented inside the human body to continuously control and monitor health conditions or a structural monitoring application having sensors deployed inside the building structures. Even in scenarios where battery replacement is not that difficult, involving humans to keep the things operational contradicts one of the main objectives of IoT: operating without any human intervention. Negative environmental impacts caused by improper disposal of a large number of depleted batteries is another detriment which reinforces the tendency towards a greener solution for supplying the energy of future IoT objects. These concerns therefore have aroused wide interests in the adoption of WPCNs into IoT and also motivated the research in this book.

References

1. N. Tesla, "Method of Regulating Apparatus For Producing Currents of High Frequency," *U.S. Patent No. 568,178*, September 1896.
2. N. Tesla, "Apparatus for Transmitting Electrical Energy,"*U.S. Patent No. 1,119,732*, December 1914.
3. W. C. Brown, " Experiments Involving a Microwave Beam to Power and Position a Helicopter," *IEEE Transactions on Aerospace and Electronic Systems*, vol. AES-5, no. 5, pp. 692–702, September 1969.
4. W. C. Brown, " The History of Power Transmission by Radio Waves," *IEEE Transactions on Microwave Theory and Techniques*, vol. 32, no. 9, pp. 1230–1242, September 1984.
5. C-S. Wang, G. A. Covic, and O. H. Stielau, "Power Transfer Capability and Bifurcation Phenomena of Loosely Coupled Inductive Power Transfer Systems," *IEEE Transactions on Industrial Electronics*, vol. 51, no. 1, pp. 148–157, February 2004.

6. B. L. Cannon, J. F. Hoburg, D. D. Stancil, and S. C. Goldstein "Magnetic Resonant Coupling As a Potential Means for Wireless Power Transfer to Multiple Small Receivers," *IEEE Transactions on Power Electronics*, vol. 24, no. 7, pp. 1819–1825, July 2009.

7. X. Lu, P. Wang, D. Niyato, D. I. Kim, and Z. Han, "Wireless Charging Technologies: Fundamentals,Standards, and Network Applications," *IEEE Communications Surveys and Tutorials*, vol. 18, no. 2, pp. 1413–1452, Second Quarter 2016.

8. S. D. Barman, A. W. Reza, N. Kumar, M. E. Karim, A. B. Munir, "Wireless powering by magnetic resonant coupling: Recent trends in wireless power transfer system and its applications," *Renewable and Sustainable Energy Reviews*, vol. 51, pp. 1525–1552, November 2015.

9. X. Lu, P. Wang, D. Niyato, and Z. Han, "Resource Allocation in Wireless Networks with RF Energy Harvesting and Transfer," *IEEE Network*, vol. 29, no. 6, pp. 68–75, November-December 2015.

10. H. T. Friis, "A Note on a Simple Transmission Formula," *Proceedings of the IRE*, vol. 34, no. 5, pp. 254–256, May 1946.

11. X. Lu, P. Wang, D. Niyato, D. I. Kim, and Z. Han, "Wireless Networks With RF Energy Harvesting: A Contemporary Survey," *IEEE Communications Surveys and Tutorials*, vol. 17, no. 2, pp. 757–789, Second Quarter 2015.

12. D. Mishra, S. De, S. Jana, S. Basagni, K. Chowdhury, and W. Heinzelman, "Smart RF Energy Harvesting Communications: Challenges and Opportunities," *IEEE Communications Magazine*, vol. 53, no. 4, pp. 70–78, April 2015.

13. M. Pinuela, P. D. Mitcheson, and S. Lucyszyn, "Ambient RF Energy Harvesting in Urban and Semi-Urban Environments," *IEEE Transactions on Microwave Theory and Techniques*, vol. 61, no. 7, pp. 2715–2726, July 2013.

14. A. Ghazanfari, H. Tabassum, and E. Hossain, "Ambient RF Energy Harvesting in Ultra-Dense Small Cell Networks: Performance and Trade-offs," *IEEE Wireless Communications*, vol. 23, no. 2, pp. 38–45, April 2016.

15. S. Lee, R. Zhang, and K. Huang, "Opportunistic Wireless Energy Harvesting in Cognitive Radio Networks," *IEEE Transactions on Wireless Communications*, vol. 12, no. 9, pp. 4788–4799, September 2013.

16. D. T. Hoang, D. Niyato, P. Wang, and D. I. Kim, "Opportunistic Channel Access and RF Energy Harvesting in Cognitive Radio Networks," *IEEE Journal on Selected Areas in Communications*, vol. 32, no. 11, pp. 2039–2052, November 2014.

17. Powercast, [Online] Available: www.powercastco.com

18. I. Krikidis, S. Timotheou, S. Nikolaou, G. Zheng, D. W. K. Ng, and R. Schober, "Simultaneous Wireless Information and Power Transfer in Modern Communication Systems," *IEEE Communications Magazine*, vol. 52, no. 11, pp. 104–110, November 2014.

19. L. R. Varshney, "Transporting Information and Energy Simultaneously," *IEEE International Symposium on Information Theory (ISIT)*, pp. 1612–1616, July 2008.

20. P. Grover and A. Sahai, "Shannon meets Tesla: Wireless information and power transfer," *IEEE International Symposium on Information Theory (ISIT)*, June 2010, pp. 2363–2367.

21. A. M. Fouladgar and O. Simeone, "On the Transfer of Information and Energy in Multi-User Systems," *IEEE Communications Letters*, vol. 16, no. 11, pp. 1733–1736, November 2012.

22. R. Zhang and C. K. Ho, "MIMO Broadcasting for Simultaneous Wireless Information and Power Transfer," *IEEE Transactions on Wireless Communications*, vol. 12, no. 5, pp. 1989–2001, May 2013.

23. S. Bi, C. K. Ho, and R. Zhang, "Wireless Powered Communication: Opportunities and Challenges," *IEEE Communications Magazine*, vol. 53, no. 4, pp. 117–125, April 2015.

24. X. Zhou, R. Zhang, and C. K. Ho, "Wireless Information and Power Transfer: Architecture Design and Rate-Energy Tradeoff," *IEEE Transactions on Communications*, vol. 61, no. 11, pp. 4754–4767, November 2013.

25. Z. Xiang and M. Tao, "Robust Beamforming for Wireless Information and Power Transmission," *IEEE Wireless Communications Letters*, vol. 1, no. 4, pp. 372–375, August 2012.
26. A. A. Nasir, X. Zhou, S. Durrani, and R. A. Kennedy, "Relaying Protocols for Wireless Energy Harvesting and Information Processing," *IEEE Transactions on Wireless Communications*, vol. 12, no. 7, pp. 3622–3636, July 2013.
27. H. Lee, C. Song, S-H. Choi, and I. Lee, "Outage Probability Analysis and Power Splitter Designs for SWIPT Relaying Systems with Direct Link," *IEEE Communications Letters*, 2016.
28. K. Huang and E. Larsson, "Simultaneous Information and Power Transfer for Broadband Wireless Systems," *IEEE Transactions on Signal Processing*, vol. 61, no. 23, pp. 5972–5986, December 2013.
29. I. Krikidis, "Simultaneous Information and Energy Transfer in Large-Scale Networks with/without Relaying," *IEEE Transactions on Communications*, vol. 62, no. 3, pp. 900–912, March 2014.
30. I. Krikidis, S. Sasaki, S. Timotheou, and Z. Ding, "A Low Complexity Antenna Switching for Joint Wireless Information and Energy Transfer in MIMO Relay Channels," *IEEE Transactions on Communications*, vol. 62, no. 5, pp. 1577–1587, May 2014.
31. C-F. Liu, M. Maso, S. Lakshminarayana, C-H. Lee, T. Q. S. Quek, "Simultaneous Wireless Information and Power Transfer Under Different CSI Acquisition Schemes,"*IEEE Transactions on Wireless Communications*, vol. 14, no. 4, pp. 1911–1926, April 2015.
32. L. Mohjazi, I. Ahmed, S. Muhaidat, M. Dianati, and M. Al-Qutayri, "Downlink Beamforming for SWIPT Multi-User MISO Underlay Cognitive Radio Networks," *IEEE Communications Letters*, 2016.
33. H. Ju and R. Zhang, "Throughput Maximization in Wireless Powered Communication Networks," *IEEE Transactions on Wireless Communications*, vol. 13, no. 1, pp. 418–428, January 2014.
34. H. Ju and R. Zhang, "Optimal Resource Allocation in Full-Duplex Wireless-Powered Communication Network," *IEEE Transactions on Communications*, vol. 62, no. 10, pp. 3528–3540, October 2014.
35. X. Kang, C. K. Ho, and S. Sun, "Full-Duplex Wireless-Powered Communication Network With Energy Causality," *IEEE Transactions on Wireless Communications*, vol. 14, no. 10, pp. 5539–5551, October 2015.
36. A. Sabharwal, P. Schniter, D. Guo, D. W. Bliss, S. Rangarajan, and R. Wichman, "In-Band Full-Duplex Wireless: Challenges and Opportunities," *IEEE Journal on Selected Areas in Communications*, vol. 32, no. 9, pp. 1637–1652, September 2014.
37. H. Ju, K. Chang, and M-S. Lee, "In-Band Full-Duplex Wireless Powered Communication Networks," *17th International Conference on Advanced Communication Technology (ICACT)*, pp. 23–27, December 2014.
38. L. Liu, R. Zhang, and K-C. Chua, "Multi-Antenna Wireless Powered Communication With Energy Beamforming," *IEEE Transactions on Communications*, vol. 62, no. 12, pp. 4349–4361, December 2014.
39. G. Yang, C. K. Ho, R. Zhang, and Y. L. Guan, "Throughput Optimization for Massive MIMO Systems Powered by Wireless Energy Transfer," *IEEE Journal on Selected Areas in Communications*, vol. 33, no. 8, pp. 1640–1650, August 2015.
40. D. Hwang, D. I. Kim, and T-J. Lee, "Throughput Maximization for Multiuser MIMO Wireless Powered Communication Networks," *IEEE Transactions on Vehicular Technology*, vol. 65, no. 7, pp. 5743–5748, July 2016.
41. D. Mishra, S. De, S. Jana, S. Basagni, K. Chowdhury, and W. Heinzelman, "Smart RF Energy Harvesting Communications: Challenges and Opportunities," *IEEE Communications Magazine*, vol. 53, no. 4, pp. 70–78, April 2015.
42. Y. L. Che, L. Duan, and R. Zhang, "Spatial Throughput Maximization of Wireless Powered Communication Networks," *IEEE Journal on Selected Areas in Communications*, vol. 33, no. 8, pp. 1534–1548, August 2015.

43. S. Lee and R. Zhang, "Cognitive Wireless Powered Network: Spectrum Sharing Models and Throughput Maximization," *IEEE Transactions on Cognitive Communications and Networking*, vol. 1, no. 3, pp. 335–346, September 2015.

44. S. S. Kalamkar, J. P. Jeyaraj, A. Banerjee, and K. Rajawat, "Resource Allocation and Fairness in Wireless Powered Cooperative Cognitive Radio Networks," *IEEE Transactions on Communications*, vol. 64, no. 8, pp. 3246–3261, August 2016.

45. Q. Wu, M. Tao, D. W. K. Ng, W. Chen, and R. Schober, "Energy-Efficient Resource Allocation for Wireless Powered Communication Networks," *IEEE Transactions on Wireless Communications*, vol. 15, no. 3, pp. 2312–2327, March 2016.

46. S. Bi, Y. Zeng, and R. Zhang, "Wireless Powered Communication Networks: An Overview," *IEEE Wireless Communications*, vol. 23, no. 2, pp. 10–18, April 2016.

47. H. Ju and R. Zhang, "User Cooperation in Wireless Powered Communication Networks," *IEEE Global Communications Conference*, pp. 1430–1435, December 2014.

48. H. Chen, Y. Li, J. L. Rebelatto, B. F. Uchoa-Filho, and B. Vucetic, "Harvest-Then-Cooperate: Wireless-Powered Cooperative Communications," *IEEE Transactions on Signal Processing*, vol. 63, no. 7, pp. 1700–1711, April 2015.

49. A. Wyner, "The wire-tap channel," *Bell Sys. Tech. J.*, vol. 54, no. 8, pp. 1355–1387, Oct. 1975.

50. D. Ng, E. Lo, and R. Schober, "Robust Beamforming for Secure Communication in Systems With Wireless Information and Power Transfer," *IEEE Trans. Wireless Commun.*, vol. 13, no. 8, pp. 4599–4615, Aug. 2014.

51. H. Zhang, C. Li, Y. Huang, and L. Yang, "Secure Beamforming for SWIPT in Multiuser MISO Broadcast Channel With Confidential Messages," *IEEE Commun. Lett.*, vol. 19, no. 8, pp. 1347–1350, Aug. 2015.

52. H. Zhang, Y. Huang, C. Li, and L. Yang, "Secure Beamforming Design for SWIPT in MISO Broadcast Channel With Confidential Messages and External Eavesdroppers," *IEEE Trans. Wireless Commun.*, vol. 15, no. 11, pp. 7807–7819, Nov. 2016.

53. F. Zhou, Z. Li, J. Cheng, Q. Li, and J. Si, "Robust AN-Aided Beamforming and Power Splitting Design for Secure MISO Cognitive Radio With SWIPT," *arXiv*, Feb. 2016. [Online]. Available: http://arxiv.org/abs/1602.06913

54. G. Pan, C. Tang, T. Li, and Y. Chen, "Secrecy Performance Analysis for SIMO Simultaneous Wireless Information and Power Transfer Systems," *IEEE Trans. Commun.*, vol. 63, no. 9, pp. 3423–3433, Sep. 2015.

55. Q. Shi, W. Xu, J. Wu, E. Song, and Y. Wang, "Secure Beamforming for MIMO Broadcasting With Wireless Information and Power Transfer," *IEEE Trans. Wireless Commun.*, vol. 14, no. 5, pp. 2841–2853, May 2015.

56. K. Banawan and S. Ulukus, "MIMO Wiretap Channel Under Receiver-Side Power Constraints With Applications to Wireless Power Transfer and Cognitive Radio," *IEEE Trans. Commun.*, vol. 64, no. 9, pp. 3872–3885, Sep. 2016.

57. W. Mou, Y. Cai, W. Yang, W. Yang, X. Xu, and J. Hu, "Exploiting full Duplex techniques for secure communication in SWIPT system," in *Proc. WCSP*, Nanjing, China, Oct. 2015, pp. 1–6.

58. W. Yang, W. Mou, X. Xu, W. Yang, and Y. Cai, "Energy efficiency analysis and enhancement for secure transmission in SWIPT systems exploiting full duplex techniques," *IET Communications*, vol. 10, no. 14, pp. 1712–1720, 2016.

59. S. S. Kalamkar and A. Banerjee, "Secure Communication Via a Wireless Energy Harvesting Untrusted Relay," *IEEE Trans. Veh. Technol.*, vol. PP, no. 99, pp. 1–1, 2016.

60. A. Salem, K. A. Hamdi, and K. M. Rabie, "Physical Layer Security With RF Energy Harvesting in AF Multi-Antenna Relaying Networks," *IEEE Trans. Commun.*, vol. 64, no. 7, pp. 3025–3038, Jul. 2016.

61. B. Li, Z. Fei, and H. Chen, "Robust Artificial Noise-Aided Secure Beamforming in Wireless-Powered Non-Regenerative Relay Networks," *IEEE Access*, vol. 4, pp. 7921–7929, 2016.

62. M. Zhao, X. Wang, and S. Feng, "Joint Power Splitting and Secure Beamforming Design in the Multiple Non-Regenerative Wireless-Powered Relay Networks," *IEEE Commun. Lett.*, vol. 19, no. 9, pp. 1540–1543, Sep. 2015.

63. H. Xing, K. K. Wong, A. Nallanathan, and R. Zhang, "Wireless Powered Cooperative Jamming for Secrecy Multi-AF Relaying Networks,"*IEEE Trans. Wireless Commun.*, vol. 15, no. 12, pp. 7971–7984, Dec. 2016.

64. N. P. Nguyen, T. Q. Duong, H. Q. Ngo, Z. Hadzi-Velkov, and L. Shu, "Secure 5g Wireless Communications: A Joint Relay Selection and Wireless Power Transfer Approach," *IEEE Access*, vol. 4, pp. 3349–3359, 2016.

65. X. Chen, J. Chen, and T. Liu, "Secure Transmission in Wireless Powered Massive MIMO Relaying Systems: Performance Analysis and Optimization," *IEEE Trans. Veh. Technol.*, vol. 65, no. 10, pp. 8025–8035, Oct. 2016.

66. L. Jiang, H. Tian, C. Qin, S. Gjessing, and Y. Zhang, "Secure Beamforming in Wireless-Powered Cooperative Cognitive Radio Networks," *IEEE Commun. Lett.*, vol. 20, no. 3, pp. 522–525, Mar. 2016.

67. W. Liu, X. Zhou, S. Durrani, and P. Popovski, "Secure Communication with a Wireless-Powered Friendly Jammer," *IEEE Trans. Wireless Commun.*, vol. 15, no. 1, pp. 401–415, Jan. 2016.

68. J. Moon, H. Lee, C. Song, and I. Lee, "Secrecy Performance Optimization for Wireless Powered Communication Networks with an Energy Harvesting Jammer," *IEEE Trans. Commun.*, vol. PP, no. 99, pp.1–1, 2016.

69. A. E. Shafie, D. Niyato, and N. Al-Dhahir, "Security of an Ordered-Based Distributive Jamming Scheme," *IEEE Communications Letters*, vol. PP, no. 99, pp. 1–1, 2016.

70. L. Tang and Q. Li, "Wireless Power Transfer and Cooperative Jamming for Secrecy Throughput Maximization," *IEEE Wireless Commun. Lett.*, vol. PP, no. 99, pp. 1–1, 2016.

71. X. Jiang, C. Zhong, Z. Zhang, and G. K. Karagiannidis, "Power Beacon Assisted Wiretap Channels With Jamming," *IEEE Trans. Wireless Commun.*, vol. 15, no. 12, pp. 8353–8367, Dec. 2016.

72. X. Jiang, C. Zhong, X. Chen, T. Q. Duong, T. A. Tsiftsis, and Z. Zhang, "Secrecy Performance of Wirelessly Powered Wiretap Channels," *IEEE Trans. Commun.*, vol. 64, no. 9, pp. 3858–3871, Sep. 2016.

73. H. Xing, K.-K. Wong, Z. Chu, and A. Nallanathan, "To Harvest and Jam: A Paradigm of Self-Sustaining Friendly Jammers for Secure AF Relaying," *IEEE Trans. Signal Process.*, vol. 63, no. 24, pp. 6616–6631, Dec. 2015.

74. A. E. Shafie, D. Niyato, and N. Al-Dhahir, "Artificial-Noise-Aided Secure MIMO Full-Duplex Relay Channels With Fixed-Power Transmissions," *IEEE Commun. Lett.*, vol. 20, no. 8, pp. 1591–1594, Aug. 2016.

75. R. Minerva, A. Biru, and D. Rotondi, "Towards a Definition of the Internet of Things (IoT)," May 2015, [online] Available: http://iot.ieee.org/definition.html.

76. H. Sundmaeker, P. Guillemin, P. Friess, and S. Woelffle, "Vision and Challenges for Realising the Internet of Things," Cluster of European Research Projects on the Internet of Things, March 2010.

77. A. McEwen and H. Cassimally, "Designing the Internet of Things," John Wiley and Sons, 2014.

78. A. Al-Fuqaha, M. Guizani, M. Mohammadi, M. Aledhari, and M. Ayyash, "Internet of Things: A Survey on Enabling Technologies, Protocols, and Applications," *IEEE Communications Surveys and Tutorials*, vol. 13, no. 4, pp. 2347–2376, Fourth Quarter 2015.

79. L. Tan and N. Wang, "Future Internet: The Internet of Things," *3rd International Conference on Advanced Computer Theory and Engineering(ICACTE)*, pp. V5-376 - V5-380, August 2010.

80. W. Ye, J. Heidemann, and D. Estrin, "An Energy-Efficient MAC Protocol for Wireless Sensor Networks," *Proceedings of the Twenty-First Annual Joint Conference of the IEEE Computer and Communications Societies*, June 2002, pp. 1567–1576.

81. T. Van Dam and K. Langendoen, "An Adaptive Energy-Efficient MAC Protocol for Wireless Sensor Networks," *Proceedings of the 1st International Conference on Embedded Networked Sensor Systems*, November 2003, pp. 171–180.

82. A. El-Hoiydi and J-D. Decotignie, "WiseMAC: An Ultra Low Power MAC Protocol for Multi-hop Wireless Sensor Networks," *Proceedings of the First International Workshop, ALGOSENSORS*, July 2004, pp. 18–31.
83. J. Polastre, J. Hill, and D. Culler, "Versatile Low Power Media Access for Wireless Sensor Networks," *Proceedings of the 2nd International Conference on Embedded Networked Sensor Systems*, November 2004, pp. 95–107.
84. Y. Yu, R. Govindan, and D. Estrin, "Geographical and Energy Aware Routing: a recursive data dissemination protocol for wireless sensor networks," *Technical Report UCLA/CSD-TR-01-0023*, UCLA Computer Science Department, May 2001.
85. D. Ganesan, R. Govindan, S. Shenker, and D. Estrin, "Highly-Resilient, Energy-Efficient Multipath Routing in Wireless Sensor Networks," *ACM SIGMOBILE Mobile Computing and Communications Review*, vol. 5, no. 4, pp. 11–25, October 2001.

Chapter 2
Enhancing Physical Layer Security in Wireless Powered Communication Networks

Abstract This chapter starts with investigating the problem of secure transmission between a wireless-powered transmitter and a receiver in the presence of multiple eavesdroppers. To counteract eavesdropping, a transmission protocol named *accumulate-then-transmit* (ATT) is proposed. Specifically, the proposed protocol employs a multi-antenna power beacon (PB) to assist the transmitter with secure transmission. First, the PB transfers wireless power to charge the transmitter's battery. After accumulating enough energy, the transmitter sends confidential information to the receiver, and simultaneously, the PB emits jamming signals (i.e., artificial noise) to interfere with the eavesdroppers. A key element of the protocol is the perfect CSI, with which the jamming signals can be deliberately designed to avoid disturbing the intended receiver. Based on the assumption that the eavesdroppers do not collude, the secrecy performance of the proposed protocol is evaluated in terms of secrecy outage probability and secrecy throughput. This study reveals that CJ is a critical enabler of PLS in WPCNs. After investigating the use of a multi-antenna PB with perfect CSI, we exploit the employment of a wireless-powered FD jammer to enhance the secrecy in the presence of CSI errors. Noteworthy, due to imperfect CSI, the jamming signals transmitted by the jammer yield undesired interference at the receiver. This study analyzes the impact of channel estimation error on the secrecy performance. Besides, due to the FD capability, the jammer is able to perform simultaneous jamming and energy harvesting. It hence makes the energy storage of the jammer experience concurrent charging and discharging. A hybrid energy storage system with finite capacity is adopted, and its long-term stationary distribution of the energy state is characterized through a finite-state Markov Chain. The secrecy performance of the proposed *accumulate-and-jam* (AnJ) protocol is evaluated to reveal its merits. Moreover, an alternative energy storage model with infinite capacity and the use of a wireless-powered HD jammer are also exploited to serve as benchmarks.

© Springer Nature Switzerland AG 2019   25
A. Jamalipour, Y. Bi, *Wireless Powered Communication Networks*,
https://doi.org/10.1007/978-3-319-98174-1_2

2.1 Accumulate-Then-Transmit: Secure WPCN in the Presence of Multiple Eavesdroppers

2.1.1 System Model and Protocol Design

As illustrated in Fig. 2.1, the system model includes a source node, Alice, powered by a dedicated PB, intends to transmit confidential information to a destination node, Bob, in the presence of N neighbor nodes. Due to size, cost, or hardware limitations, all wireless nodes except PB may not be able to support multiple transmit antennas, e.g., handsets and sensors. As a result, in this chapter, it is assumed that only PB has $K > 1$ antennas, whereas all other nodes are each equipped with a single antenna. With the single antenna setup, the confidential information sent by Alice is accessible by not only Bob but also other neighbor nodes. Out of curiosity, these neighbors, although assumed non-colluding, may play the roles of potential eavesdroppers by intercepting information that are not intended for them to decode.

All wireless channels in the considered system model are assumed as quasi-static and flat fading. Fading channel coefficients thereby remain constant within each transmission block of duration T and vary independently from one block to another. Let $\mathbf{g}_{pa}$ and h_{ab} denote the channel coefficients of the links PB $\rightarrow$ Alice and Alice $\rightarrow$ Bob, respectively. The entries of $\mathbf{g}_{pa}$ and h_{ab} are assumed to be independent and identically distributed (i.i.d.) zero mean complex Gaussian variables with variance λ_{pa} and λ_{ab}, respectively. Also, denote the set of all eavesdroppers as Φ_E. Let $\mathbf{g}_{pe_n}$ and h_{ae_n} denote the respective channel coefficients for the channel from PB and Alice to the n-th (i.e., $j \in \Phi_E$) eavesdropper. The entries of $\mathbf{g}_{pe_n}$ and h_{ae_n} are assumed to be i.i.d. zero mean complex Gaussian variables with variances λ_{ge} and λ_{ae}, respectively. To capture the effect of large-scale path loss on the network performance, the bounded path-loss model[1] is adopted to define the channel

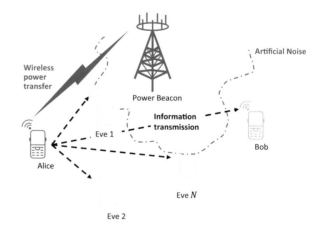

Fig. 2.1 A schematic diagram of the proposed accumulate-then-transmit protocol

---

[1]The considered bounded path-loss model ensures that the path loss is always larger than 1 for any distance and has been widely adopted in the literature [13–16].

variance as $\lambda_{XY} = 1/(1 + d_{XY}^{\delta})$; d_{XY} denotes the distance between nodes X and Y and $\delta \in [2, 5]$ is the path-loss factor. For all the wireless channels in the considered system, channel reciprocity is assumed. It is also important to note that since the eavesdroppers are passive, i.e., they only listen to the transmission of Alice, the instantaneous CSI of the N eavesdroppers are unknown at other nodes.[2]

2.1.1.1 Performance Metrics

Since the CSI of the eavesdroppers is unknown, Alice uses Wyner coding for information transmission [18]. Specifically, Alice chooses two constant rates, namely the codeword transmission rate R_t and the secrecy transmission rate R_s. The rate redundancy, i.e. $R_e = R_t - R_s$, reflects the coding cost of securing the confidential message against intercepting. Based on that, the following three performance metrics are defined:

- **Connection Outage**: The connection outage is defined as the probability that the capacity of the channel from Alice to Bob is below the codeword transmission rate R_t.
- **Secrecy Outage**: The secrecy outage is defined as the probability that the maximal capacity of the channels from Alice to multiple eavesdroppers is above the rate redundancy R_e.
- **Secrecy Throughput**: With a certain connection outage probability p_{co} and secrecy outage probability p_{so}, the secrecy throughput ξ_{st} is defined as

$$\xi_{st} = (1 - p_{co})(1 - p_{so})R_s. \tag{2.1.1}$$

With the selected Wyner code (R_t, R_s), the probability of the reliable transmission is $1 - p_{co}$, and the probability that the transmitted signal being secured against multiple eavesdroppers is $1 - p_{so}$.

2.1.1.2 Protocol Description

At the beginning of each transmission block, Bob sends Alice a pilot signal for channel estimation, e.g., using channel training methods designed specifically for WPCNs [19]. With the assumed channel reciprocity, Alice can estimate h_{ab}, which is further used to calculate the minimum transmit power P_a required for guaranteeing a connection outage-free transmission. Mathematically, P_a and h_{ab} should satisfy

---

[2]But it is assumed that the channel distributions of eavesdroppers and their channel variances are available. This assumption has been widely adopted in the literature [1, 9, 17].

$$\log_2 \left(1 + \frac{P_a |h_{ab}|^2}{\sigma_b^2} \right) = R_t, \qquad (2.1.2)$$

where σ_b^2 is the variance of the AWGN at Bob. After some simple manipulation, we obtain

$$P_a = \frac{(2^{R_t} - 1)\sigma_b^2}{H_{ab}}, \qquad (2.1.3)$$

where $H_{ab} := |h_{ab}|^2$ is the channel gain of the link Alice-Bob.

Next, Alice verifies whether her residual energy E_a is sufficient for a connection outage-free transmission. If $E_a \geq P_a T$, Alice notifies Bob and PB by broadcasting one single bit "1" that the information transmission (IT) mode should be invoked. Otherwise, if Alice has insufficient energy, i.e., $E_a < P_a T$, Alice sends Bob and PB one single bit "0" to activate the wireless power transfer (WPT) mode. In the following, the details of the signal processing occurred in each mode are presented.

In the WPT mode, with a fixed transmit power P_0, PB sends Alice an energy-bearing signal which is randomly generated and contains no secret information. To maximize energy acquisition, PB implements maximum ratio transmission (MRT) with beamforming vector $\mathbf{w} = \mathbf{g}_{pa}/\|\mathbf{g}_{pa}\|$. By ignoring the negligible energy harvested from the noise, the amount of acquired energy by Alice can be expressed as

$$E_h = \eta P_0 G_{pa} T, \qquad (2.1.4)$$

where $G_{pa} = \|\mathbf{g}_{pa}\|^2$ denotes the channel gain of PB-Alice link, and $0 < \eta < 1$ is the energy conversion efficiency of the energy harvesting circuit of Alice [20, 21].

In the IT mode, Alice uses the energy harvested in the WPT mode to transmit the information signal x_a to Bob, under the protection of the jamming signal $\mathbf{x}_j$ sent by PB. PB uses the same power level P_0 for both energy transfer and signal transmission. It is worth noting that, due to the lack of the eavesdroppers' instantaneous CSI, finding the optimal $\mathbf{x}_j$ is infeasible. To tackle the problem, the artificial interference generation method is used to generate a suboptimal jamming signal, such that the yielded interference confounds only the potential eavesdroppers but not Bob [5, 7]. Specifically, the jamming signal is in the form of $\mathbf{x}_j = \mathbf{Wv}$, where the $K \times (K - 1)$ matrix $\mathbf{W}$ is an orthogonal basis of the null space of the link PB $\rightarrow$ Bob, and the vector $\mathbf{v}$ has $K - 1$ i.i.d. complex Gaussian random elements with unit variance. The jamming power of PB uniformly distributes among the $K - 1$ dimensions.

Based on the design of $\mathbf{x}_j$, the received signals at Bob and at the n-th eavesdropper are given by

$$y_b = \sqrt{P_a} h_{ab} x_a + n_b, \qquad (2.1.5)$$

and

$$y_{e_n} = \sqrt{P_a} h_{ae_n} x_a + \sqrt{P_0} \mathbf{g}_{pe_n}^\dagger \frac{\mathbf{Wv}}{K-1} + n_e, \qquad (2.1.6)$$

where $h_{ab}, h_{ae_n} \in \mathbb{C}^{1\times 1}$ and $\mathbf{g}_{pe_n} \in \mathbb{C}^{K\times 1}$. It is worth noting that h_{ae_n} and $\mathbf{g}_{pe_n}$ are unknown by Alice, Bob, and the PB due to the assumption of passive eavesdropping. n_b and n_e are the AWGN at Bob and the n-th eavesdropper, with the variance σ_b^2 and σ_e^2, respectively.

2.1.2 Battery State Analysis

Since a practical energy storage with finite capacity is considered in this work, the analyses designed for infinite battery capacity [9] are not applicable to the current study. Therefore, we follow [12] and apply a discrete-level battery model to characterize the dynamic behaviors of Alice's battery. Specifically, we discretize the battery storage into $L+1$ energy levels, with the i-th level expressed as $\varepsilon_i = iC/L$, $i \in \{0, 1, \ldots, L\}$, with C representing the battery capacity. Specifically, in the WPT mode, the discretized energy saved in the battery can be expressed as

$$\varepsilon_h = \varepsilon_j, \text{ where } j = \arg \max_{i \in \{0,1,\ldots,L\}} \left\{ \varepsilon_i : \varepsilon_i \leq E_h \right\}. \qquad (2.1.7)$$

It is worth pointing out that if $E_h > C$, energy overflow will happen because the maximum amount of energy that can be saved at Alice is C. Equation (2.1.7) implies $\varepsilon_h \leq \varepsilon_L = C$ by limiting $i \in \{0, 1, \ldots, L\}$.

On the other hand, in the IT mode, the consumed energy, i.e., $P_a T$, corresponds to a discrete energy level ε_t, which is defined as

$$\varepsilon_t = \varepsilon_j, \text{ where } j = \arg \min_{i \in \{0,1,\ldots,L\}} \left\{ \varepsilon_i : \varepsilon_i \geq P_a T \right\}. \qquad (2.1.8)$$

Based on the principle of the proposed ATT protocol, we can now readily describe how the residual energy at Alice evolves along the communication process. Let $\zeta[t] \in \{\zeta_{WPT}, \zeta_{IT}\}$ denote the operation mode of Alice for the t-th transmission block, where ζ_{WPT} and ζ_{IT} represent the WPT and IT modes, respectively. Then, $\zeta[t]$ is determined by Alice's residual energy $\varepsilon_r[t]$ at the beginning of that transmission block. Mathematically, we have

$$\zeta[t] = \begin{cases} \zeta_{IT} & \text{if } \varepsilon_r[t] \geq \varepsilon_t. \\ \zeta_{WPT} & \text{otherwise} \end{cases} \qquad (2.1.9)$$

Further, the energy evolution at Alice can be described as follows:

$$\varepsilon_r[t+1] = \begin{cases} \min\left\{\varepsilon_r[t] + \varepsilon_h, \ C\right\} & \text{if } \zeta[t] = \zeta_{WPT}, \\ \varepsilon_r[t] - \varepsilon_t & \text{if } \zeta[t] = \zeta_{IT}. \end{cases} \tag{2.1.10}$$

It is worth noting that the discrete energy model can tightly approximate its continuous counterpart when the number of the discretization level is sufficiently large (i.e., $L \to \infty$ corresponds to a continuous energy storage model).

2.1.2.1   Markov Chain

With the energy discretization described above, the transition of Alice's energy states can be modeled as a finite-state MC. Define state S_i as the residual energy of Alice ε_r being equal to ε_i. Next, we derive the probability $p_{i,j}$ of a transition from the state S_i to the state S_j as follows.

1. *The battery remains empty* $(S_0 \to S_0)$: This transition case starts with the state S_0, i.e., the battery is empty. In this case, the WPT mode should be invoked. The case that the transition finishes with the same state S_0 indicates that after discretization, the harvested energy becomes zero, i.e., $\varepsilon_h = 0$. According to the definition in (2.1.7), the condition $E_h < C/L$ must hold. Hence, the transition probability of this case can be expressed as

$$p_{0,0} = \Pr\left\{\varepsilon_h = 0\right\}$$

$$= \Pr\left\{G_{pa} < \frac{C}{\eta P_0 T L}\right\}$$

$$= 1 - \exp\left(-\frac{C}{\eta \lambda_{pa} P_0 T L}\right) \sum_{k=0}^{K-1} \frac{\left(\frac{C}{\eta \lambda_{pa} P_0 T L}\right)^k}{k!}. \tag{2.1.11}$$

where (2.1.11) is obtained based on the fact that G_{pa} is gamma-distributed with parameters (K, λ_{pa}).

2. *The battery is fully charged* $(S_0 \to S_L)$: The empty energy state indicates that the WPT mode is activated. To make the battery become fully charged, the amount of harvested energy should be no less than the capacity of Alice's rechargeable battery. The transition probability is hence given by

$$p_{0,L} = \Pr\left\{\varepsilon_h \geq C\right\}$$

$$= \Pr\left\{G_{pa} \geq \frac{C}{\eta P_0 T}\right\}$$

$$= \exp\left(-\frac{C}{\eta \lambda_{pa} P_0 T}\right) \sum_{k=0}^{K-1} \frac{\left(\frac{C}{\eta \lambda_{pa} P_0 T}\right)^k}{k!}. \tag{2.1.12}$$

3. *The battery is partially charged* $(S_0 \rightarrow S_i : 0 < i < L)$: The state transition indicates that the discretized amount of harvested energy ε_h should equal iC/L. Therefore, E_h must fall between the energy level iC/L and $(i + 1)C/L$. The transition probability is thus calculated as

$$p_{0,i} = \Pr\left\{\varepsilon_h = iC/L\right\}$$

$$= \Pr\left\{\frac{iC}{\eta P_0 T L} \leq G_{pa} < \frac{(i+1)C}{\eta P_0 T L}\right\}$$

$$= \exp\left(-\frac{iC}{\eta \lambda_{pa} P_0 T L}\right) \sum_{k=0}^{K-1} \frac{\left(\frac{iC}{\eta \lambda_{pa} P_0 T L}\right)^k}{k!}$$

$$- \exp\left(-\frac{(i+1)C}{\eta \lambda_{pa} P_0 T L}\right) \sum_{k=0}^{K-1} \frac{\left(\frac{(i+1)C}{\eta \lambda_{pa} P_0 T L}\right)^k}{k!}. \tag{2.1.13}$$

4. *The non-empty battery is partially charged* $(S_i \rightarrow S_j : 0 < i < j < L)$: The WPT mode is selected in this transition case, and the discrete amount of harvested energy equals $(j - i)C/L$. Therefore, E_h must fall between the energy level $(j - i)C/L$ and $(j - i + 1)C/L$. The transition probability is

$$p_{i,j} = \Pr\left\{(\varepsilon_t > iC/L) \bigcap (\varepsilon_h = (j - i)C/L)\right\}$$

$$= \Pr\left\{H_{ab} < \frac{\alpha \sigma_b^2 T L}{iC}\right\} \Pr\left\{\frac{(j-i)C}{\eta P_0 T L} \leq G_{pa} < \frac{(j-i+1)C}{\eta P_0 T L}\right\}$$

$$= \left[1 - \exp\left(-\frac{\alpha \sigma_b^2 T L}{i \lambda_{ab} C}\right)\right]$$

$$\times \left[\exp\left(-\frac{(j-i)C}{\eta \lambda_{pa} P_0 T L}\right) \sum_{k=0}^{K-1} \frac{\left(\frac{(j-i)C}{\eta \lambda_{pa} P_0 T L}\right)^k}{k!}\right.$$

$$\left. - \exp\left(-\frac{(j-i+1)C}{\eta \lambda_{pa} P_0 T L}\right) \sum_{k=0}^{K-1} \frac{\left(\frac{(j-i+1)C}{\eta \lambda_{pa} P_0 T L}\right)^k}{k!}\right], \tag{2.1.14}$$

where we define $\alpha := 2^{R_t} - 1$ for notation simplicity.

5. *The non-empty and non-full battery remains unchanged* ($S_i \rightarrow S_i : 0 < i < L$): In this transition case, the WPT mode is activated, and the discrete amount of harvested energy equals zero. It requires that the minimum transmit energy ε_t is greater than the current residual energy iC/L. The transition probability is given by

$$
p_{i,i} = \Pr\left\{(\varepsilon_t > iC/L) \bigcap (\varepsilon_h = 0)\right\}
$$

$$
= \Pr\left\{H_{ab} < \frac{\alpha \sigma_b^2 T L}{iC}\right\} \Pr\left\{G_{pa} < \frac{C}{\eta P_0 T L}\right\}
$$

$$
= \left[1 - \exp\left(-\frac{\alpha \sigma_b^2 T L}{i \lambda_{ab} C}\right)\right]
$$

$$
\times \left[1 - \exp\left(-\frac{C}{\eta \lambda_{pa} P_0 T L}\right) \sum_{k=0}^{K-1} \frac{\left(\frac{C}{\eta \lambda_{pa} P_0 T L}\right)^k}{k!}\right]. \tag{2.1.15}
$$

6. *The non-empty battery is fully charged* ($S_i \rightarrow S_L : 0 < i < L$): Similar to the previous case, only the discrete harvested energy equals $(L - i)C/L$. The transition probability is thus given by

$$
p_{i,L} = \Pr\left\{(\varepsilon_t > iC/L) \bigcap (\varepsilon_h = (L - i)C/L)\right\}
$$

$$
= \Pr\left\{H_{ab} < \frac{\alpha \sigma_b^2 T L}{iC}\right\} \Pr\left\{G_{pa} \geq \frac{(L - i)C}{\eta P_0 T L}\right\}
$$

$$
= \left[1 - \exp\left(-\frac{\alpha \sigma_b^2 T L}{i \lambda_{ab} C}\right)\right] \exp\left(-\frac{(L - i)C}{\eta \lambda_{pa} P_0 T L}\right) \sum_{k=0}^{K-1} \frac{\left(\frac{(L-i)C}{\eta \lambda_{pa} P_0 T L}\right)^k}{k!}.
$$

$$
\tag{2.1.16}
$$

7. *The battery remains full* ($S_L \rightarrow S_L$): In this case, the WPT mode is performed in the current transmission block and the harvested energy can be any arbitrary value since the full battery cannot save any more energy. The transition probability is

$$
p_{L,L} = \Pr\left\{\varepsilon_t > C\right\} = \Pr\left\{H_{ab} < \frac{\alpha \sigma_b^2 T}{C}\right\} = 1 - \exp\left(-\frac{\alpha \sigma_b^2 T}{\lambda_{ab} C}\right). \tag{2.1.17}
$$

8. *The battery is discharged* ($S_j \rightarrow S_i : 0 \leq i < j \leq L$): According to the definition in (2.1.10), it can be easily inferred that the energy level reduces only when the IT

mode is invoked. The state transition from S_j to S_i indicates that the discretized amount of energy ε_t equals $(j - i)C/L$. Therefore, the consumed energy for information transmission, i.e., $P_a T$, falls between the energy level $(j-i-1)C/L$ and $(j - i)C/L$. Hence, the transition probability is given by

$$
\begin{aligned}
p_{j,i} &= \Pr\left\{\varepsilon_t = (j - i)C/L\right\} \\
&= \Pr\left\{\frac{\alpha\sigma_b^2 TL}{(j - i)C} \le H_{ab} < \frac{\alpha\sigma_b^2 TL}{(j - i - 1)C}\right\} \\
&= \exp\left(-\frac{\alpha\sigma_b^2 TL}{(j - i)\lambda_{ab}C}\right) - \exp\left(-\frac{\alpha\sigma_b^2 TL}{(j - i - 1)\lambda_{ab}C}\right).
\end{aligned}
\tag{2.1.18}
$$

The objective is to find the stationary distribution π of the MC; $\pi_i, i \in \{0, 1, \ldots, L\}$ denotes the probability that the residual energy of the battery is ε_i. Let $\mathbf{M} = (p_{i,j})$ denote the $(L + 1) \times (L + 1)$ state transition matrix of the MC. By using the similar method in [12], we can readily verify that the MC transition matrix $\mathbf{M}$ is irreducible and row stochastic. Therefore, there must exist a unique solution π that satisfies the following equation,

$$
\pi = (\pi_0, \pi_1, \ldots, \pi_L)^T = \mathbf{M}^T \pi.
\tag{2.1.19}
$$

Consequently, the stationary distribution of Alice's energy can be obtained as

$$
\pi = (\mathbf{M}^T - \mathbf{I} + \mathbf{B})^{-1}\mathbf{b},
\tag{2.1.20}
$$

where $\mathbf{B}_{i,j} = 1, \forall i, j$ and $\mathbf{b} = (1, 1, \ldots, 1)^T$.

2.1.3 Performance Evaluation

In this section, the achievable performance of the proposed secrecy transmission scheme are examined. Under a fixed R_t and R_s, the analyses of the connection outage and the secrecy outage are given in Sects. 2.1.3.1 and 2.1.3.2. In Sect. 2.1.3.3, the secrecy throughput is characterized.

2.1.3.1 Connection Outage Probability

Since Alice transmits information to Bob only in the IT mode, based on (2.1.5) the received signal-to-noise ratio (SNR) at Bob is given by

$$\gamma_b = \begin{cases} \frac{P_a |h_{ab}|^2}{\sigma_b^2} & \text{if } \zeta = \zeta_{IT}, \\ 0 & \text{if } \zeta = \zeta_{WPT}. \end{cases} \tag{2.1.21}$$

Consequently, the connection outage probability of the proposed ATT scheme is

$$p_{co} = \Pr\left\{ \log_2 \left(1 + \gamma_b\right) < R_t \right\}. \tag{2.1.22}$$

and its analytical result is given in Proposition 2.1.

Proposition 2.1 *Denoting* $\alpha = 2^{R_t} - 1$, *the connection outage probability in* (2.1.22) *of the proposed ATT scheme can be expressed as*

$$p_{co} = 1 - \sum_{i=1}^{L} \pi_i \exp\left(-\frac{\alpha \sigma_b^2 T L}{i \lambda_{ab} C} \right). \tag{2.1.23}$$

Proof Rather than directly deal with the connection outage, we instead calculate the probability of its complementary event, i.e., the transmission probability $p_{tx} = 1 - p_{co}$. By applying the total probability theorem, p_{tx} can be expressed as

$$\begin{aligned} p_{tx} &= \Pr\left\{ \log_2 \left(1 + \gamma_b\right) \geq R_t \right\} \\ &= \Pr\left\{ \log_2 \left(1 + \gamma_b\right) \geq R_t | \zeta = \zeta_{WPT} \right\} \Pr\left\{ \zeta = \zeta_{WPT} \right\} \\ &\quad + \Pr\left\{ \log_2 \left(1 + \gamma_b\right) \geq R_t | \zeta = \zeta_{IT} \right\} \Pr\left\{ \zeta = \zeta_{IT} \right\}. \end{aligned} \tag{2.1.24}$$

Based on (2.1.21), we have $\Pr\{\log_2 \left(1 + \gamma_b\right) \geq R_t | \zeta = \zeta_{WPT}\} = 0$ and $\Pr\{\log_2 \left(1 + \gamma_b\right) \geq R_t | \zeta = \zeta_{IT}\} = 1$. Therefore, we can recast p_{tx} as

$$\begin{aligned} p_{tx} &= \Pr\left\{ \zeta = \zeta_{IT} \right\} \\ &= \sum_{i=1}^{L} \Pr\left\{ (\varepsilon_r = iC/L) \bigcap (\varepsilon_t \leq iC/L) \right\} \\ &= \sum_{i=1}^{L} \pi_i \Pr\left\{ H_{ab} \geq \frac{\alpha \sigma_b^2 T L}{iC} \right\} \\ &= \sum_{i=1}^{L} \pi_i \exp\left(-\frac{\alpha \sigma_b^2 T L}{i \lambda_{ab} C} \right). \end{aligned} \tag{2.1.25}$$

Therefore, we can obtain the connection outage probability by calculating $p_{co} = 1 - p_{tx}$. This completes the proof. ∎

Remark 2.1 Equation (2.1.23) is verified by comparing the theoretical results obtained in (2.1.23) with Monte Carlo simulation results. In Monte Carlo simulations, the system parameters are set as $K = 6, T = 1, C = 0.01, L = 100, \eta = 0.5, d_{ab} = 20$ m, $\delta = 2$, and $\sigma_b^2 = -50$ dBm. It is observed that in Fig. 2.2 the theoretical curves coincide well with the simulation ones, which validates Proposition 2.1. In addition, Fig. 2.2 also shows the impact of R_t on the connection outage probability, i.e., p_{co} increases with the increasing $R_t = [1, 4, 6]$. This can be easily explained by the definition of p_{co} in (2.1.22). Furthermore, it is observed that p_{co} decreases monotonically as P_0 increases. The reason is that with the increasing P_0, Alice can harvest more energy in each transmission block which would certainly increase the transmission probability, and thus reduce the connection outage.

2.1.3.2 Secrecy Outage Probability

In this subsection, we characterize the secrecy outage probability p_{so}. From (2.1.6), the received signal-to-interference-noise ratio (SINR) at the n-th eavesdropper is

$$\gamma_{e_n} = \begin{cases} P_a |h_{ae_n}|^2 \Big/ \left(\frac{P_0}{K-1} ||\mathbf{g}_{pe_n}^\dagger \mathbf{W}||^2 + \sigma_e^2 \right) & \text{if } \zeta = \zeta_{IT}, \\ 0 & \text{if } \zeta = \zeta_{WPT}. \end{cases} \quad (2.1.26)$$

Denoting the rate redundancy threshold as $R_e = R_t - R_s$ and the corresponding channel capacity threshold for the secrecy outage as $\beta = 2^{R_e} - 1$, the corresponding secrecy outage probability is

Fig. 2.2 Connection outage probability of the proposed ATT scheme versus P_0

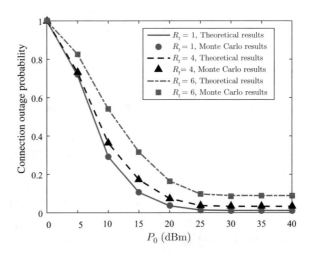

$$p_{so} = \Pr\left\{ \max_{n \in \Phi_E}\{\gamma_{e_n}\} > \beta \right\}. \tag{2.1.27}$$

In the following, the theoretical result of (2.1.27) is derived. We first formalize the following lemma to facilitate the secrecy outage probability analysis.

Lemma 2.1 *Assuming that a random variable X is exponential-distributed with the rate parameter λ_X and a random variable Y is gamma-distributed with the shape parameter k_Y and scale parameter θ_Y, then the CDF of the random variable $Z = X/(Y + a)$ is given by*

$$F_Z(z) = 1 - e^{-az/\lambda_X}\left(\frac{k_Y}{k_Y\theta_Y z/\lambda_X + k_Y}\right)^{k_Y}. \tag{2.1.28}$$

Based on Lemma 2.1, we compute the closed-form expression of p_{so} as follows.

Proposition 2.2 *For a given rate redundancy $R_e = R_t - R_s$, the secrecy outage probability of the proposed ATT scheme is given by*

$$
\begin{aligned}
p_{so} = \sum_{i=1}^{L}\sum_{m=1}^{N}\pi_i\binom{N}{m}\frac{(-1)^{m+1}}{\lambda_{ab}}\left(\frac{K-1}{\psi}\right)^{m(K-1)} \\
\times \exp\left(-\left(\phi m + \frac{1}{\lambda_{ab}}\right)\frac{\alpha\sigma_b^2 T L}{iC}\right) \\
\times \Psi\left(m(K-1), \phi m + \frac{1}{\lambda_{ab}}, \frac{\alpha\sigma_b^2 T L}{iC} + \frac{K-1}{\psi}\right).
\end{aligned} \tag{2.1.29}
$$

where for notation simplicity, we define

$$\phi := \frac{\sigma_e^2\beta}{\alpha\sigma_b^2\lambda_{ae}}, \qquad \psi := \frac{P_0\lambda_{pe}\beta}{\alpha\sigma_b^2\lambda_{ae}}. \tag{2.1.30}$$

and

$$
\begin{aligned}
\Psi(n, \mu, \nu) &:= \int_0^\infty \frac{e^{-\mu x}}{(x + \nu)^n}dx \\
&= \frac{1}{(n-1)!}\sum_{k=1}^{n-1}(k-1)!(-\mu)^{n-k-1}\nu^{-k} \\
&\quad - \frac{(-\mu)^{n-1}}{(n-1)!}e^{\mu\nu}\mathrm{Ei}(-\mu\nu)
\end{aligned} \tag{2.1.31}
$$

Proof See Appendix A. ∎

Fig. 2.3 Secrecy outage probability of the proposed ATT scheme versus P_0. The system parameters are $K = 6$, $N = 3$, $T = 1$, $C = 0.01$, $L = 100$, $\eta = 0.5$, $[d_{ab}, d_{ae}, d_{ha}, d_{he}] = [20, 10, 10, 20]$ m, $\delta = 2$, and $\sigma_b^2 = \sigma_e^2 = -50$ dBm

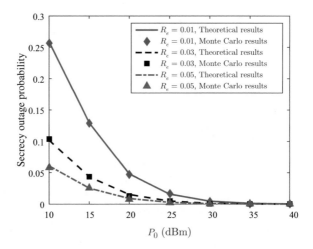

Remark 2.2 The proof of Proposition 2.2 differs from the existing works as we have taken the thermal noise at Eves into considerations. Although the noise power at Eves are typically unknown to Alice, to evaluate the secrecy performance of the proposed ATT scheme in a practical scenario, in Monte Carlo simulations we consider the noise power at Eves are as same as the noise power at Bob. Simulation results plotted in Fig. 2.3 validate Proposition 2.2. From Fig. 2.3, it is also observed that increasing P_0 and/or R_e can significantly improve the secrecy outage probability. The reason is that for a given R_s, larger R_e means that more code redundancy is used to protect the confidential message.

2.1.3.3 Secrecy Throughput

In this subsection, we characterize the secrecy throughput ξ_{st} in the following corollary.

Corollary 2.1 *Combining the derived connection outage probability in (2.1.23) with the secrecy outage probability in (2.1.29), and substituting into (2.1.1), we can obtain the closed form expression for the secrecy throughput of the proposed ATT scheme given by*

$$\xi_{st} = R_s \sum_{i=1}^{L} \pi_i \exp\left(-\frac{\alpha \sigma_b^2 T L}{i \lambda_{ab} C}\right)$$

$$\times \left\{1 - \sum_{i=1}^{L} \sum_{m=1}^{N} \pi_i \binom{N}{m} \frac{(-1)^{m+1}}{\lambda_{ab}} \left(\frac{K-1}{\psi}\right)^{m(K-1)}\right\}$$

$$\times \exp\left(-\left(\phi m + \frac{1}{\lambda_{ab}}\right) \frac{\alpha \sigma_b^2 T L}{i C}\right)$$

$$\times \Psi\left(m(K-1), \phi m + \frac{1}{\lambda_{ab}}, \frac{\alpha \sigma_b^2 T L}{i C} + \frac{K-1}{\psi}\right)\Bigg\}. \qquad (2.1.32)$$

Remark 2.3 In Remarks 2.1 and 2.2, we elaborate that the increasing R_t has a negative impact on the connection outage probability but a positive impact on the secrecy outage probability. As a result, based on (2.1.1), there should be an optimal R_t^* that can find the best trade-off between the probability of $1 - p_{co}$ and $1 - p_{so}$ for maximizing ξ_{st}. Figure 2.4 depicts the change trend of the secrecy throughput versus the increasing R_t. It shows that as R_t increases, the secrecy throughput first increases then decreases. The optimal R_t^* with the maximum secrecy throughput achieves the best trade-off between increasing the transmission probability and reducing the secrecy outage probability. Unfortunately, due to the complexity of the considered MC model, a general closed-form solution for R_t^* is difficult to derive. Nevertheless, for a given network setup with a maximum allowable connection outage probability $p_{co}^{\max}$, we can calculate the maximum allowable transmission rate $R_t^{\max}$ based on (2.1.23). Then, the optimal R_t^* can be readily obtained by performing a one-dimensional exhaustive search over its feasible range $(R_s, R_t^{\max})$.

2.1.4 Numerical Results

In this section, more representative simulation results are provided to gain more insights about the proposed ATT scheme. In line with [9], the simulation set up is

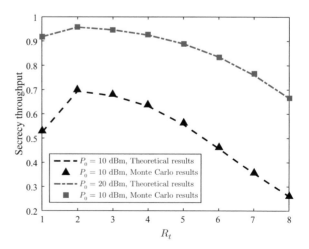

Fig. 2.4 Secrecy throughput of the proposed ATT scheme versus R_t. The system parameters are $K = 6$, $N = 3$, $T = 1$, $C = 0.01$, $L = 100$, $\eta = 0.5$, $R_s = 1$, $[d_{ab}, d_{ae}, d_{ha}, d_{he}] = [20, 10, 10, 20]$ m, $\delta = 2$, and $\sigma_b^2 = \sigma_e^2 = -50$ dBm

Fig. 2.5 Connection outage probability versus K with different Alice-PB distances

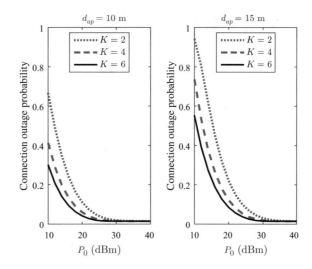

based on a linear topology where Eves, Alice, PB, and Bob are placed in order along a horizontal line; the distances are set to $d_{ea} = d_{ap} = d_{pb} = 10$ m. Throughout this section, unless otherwise stated, we normalize the transmission block duration $T = 1$, and set the pass-loss factor $\delta = 2$, the number of eavesdroppers $N = 3$, the number of antennas at PB $K = 4$, the noise power $\sigma_b^2 = \sigma_e^2 = -50$ dBm, the transmission rate $R_t = 2$, and the secrecy rate $R_s = 1$. For system parameters regarding the energy storage at Alice, we choose the energy conversion efficiency $\eta = 0.5$ [9, 22], battery capacity $C = 0.01$, and discretization levels $L = 100$. It is worth noting that since the accuracy of the theoretical results have already been validated in Sect. 2.1.3, in all the following figures we will omit Monte Carlo simulation results.

In Fig. 2.5, we evaluate the impact of the distance between PB and Alice d_{ap} as well as the number of antennas K at PB on connection outage. It is observed that as d_{ap} increases from 10 to 15 m, the connection outage probability increases notably. The reason is that by increasing the distance between PB and Alice, the amount of energy that Alice can harvest from PB reduces. In this case, according to the proposed ATT protocol, more blocks are used for the WPT mode. Consequently, more WPT blocks leads to higher connection outage probability. Additionally, in both cases of Fig. 2.5 it is observed that increasing K can significantly reduce p_{co} when P_0 is below 30 dBm. This finding suggests that in the low transmit power regime, equipping more antennas at PB is an effective method to improve network connection regardless of PB's position.

Figure 2.6 plots the effect of the number of eavesdroppers N on the secrecy outage probability. It is observed that the achievable secrecy outage probability becomes worse as N increases. Furthermore, it is notable that with the increasing P_0, the secrecy outage probability increases first and then declines. When P_0 is small, Alice cannot harvest enough energy for supporting information transmission.

Fig. 2.6 Secrecy outage
probability versus N

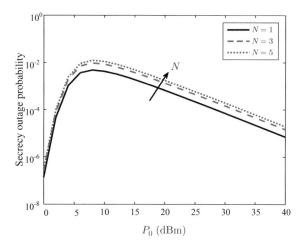

Hence, the WPT mode dominates, and the secrecy outage hardly happens. As P_0 increases, more blocks are used for the IT mode. However, since the jamming power is not adequate to yield sufficient interference at the eavesdroppers, p_{so} increases consequently. When P_0 exceeds 10 dBm in Fig. 2.6, the eavesdroppers are overwhelmed by the jamming signals, and therefore, the secrecy outage probability keeps decreasing.

Finally, we compare the achievable secrecy throughput of the proposed ATT scheme with the benchmark *harvest-then-transmit* (HTT) scheme [2, 10]. In HTT, each transmission block is divided into two phases. During the first τT $(0 < \tau < 1)$ amount of time, the transmitter harvests energy from the RF signals transmitted by PB. In the remaining $(1 - \tau)T$ amount of time, the transmitter uses the harvested energy to transmit information signal to the receiver, while PB transmits jamming signal to confound the eavesdroppers at the same time. Optimization techniques have been applied to maximize the secrecy throughput via finding the optimal time partition ratio τ.

In Fig. 2.7, it is observed that with the number of antennas at PB equal to 2, the HTT scheme can achieve larger secrecy throughput than our proposed ATT when $P_0 < 17$ dBm. The reason is that small P_0 and K in ATT scheme would lead to transmission blocks mostly used for the WPT mode, hence the IT mode can hardly take place. But in HTT scheme, a fraction of each transmission block is always allocated for information transmission regardless of the system parameter. Nevertheless, in the mid-high P_0 regime, the performance of ATT rapidly overtakes that of HTT as expected. When $K = 6$, our proposed ATT scheme always outperforms the HTT scheme. As per (2.1.1), the higher throughput of ATT is as a result of the improved connection outage and the secrecy outage. It suggests that when the number of antennas at PB is large, it is always superior to enable energy accumulation at the wireless powered nodes.

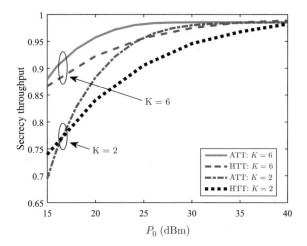

Fig. 2.7 Secrecy throughput versus P_0 achieved by the proposed ATT scheme and the benchmark HTT scheme

2.2   Accumulate-and-Jam: Secure WPCN via A Wireless-Powered Full-Duplex Jammer

2.2.1   System Model and Protocol Design

We consider a point-to-point Gaussian wiretap channel which consists of a source (S), a destination (D), and an eavesdropper (E). Each of these three wireless agents, either being a handset or sensor, is equipped with a single antenna, as illustrated in Fig. 2.8. In line with the vast majority of previous studies, the eavesdropper in this work is considered to be a passive adversary, i.e., it may not transmit but only listens. As such, the instantaneous CSI of E are unknown by any other nodes in the network.[3] Moreover, the passive eavesdropping makes the source (S) impossible to determine whether the main channel (S → D) is superior to the wiretap channel (S → E).

2.2.1.1   Jammer Model

The jammer is assumed to be an energy-constrained node without embedded energy source. It thus needs to acquire energy from ambient RF signals to function. Specifically, when performing cooperative jamming, J operates in FD mode: it uses N_r (i.e., $N_r \geq 1$) antennas to harvest energy from S, and N_t (i.e., $N_t \geq 2$) antennas to transmit jamming signals, simultaneously. When cooperative jamming is not carried out, J focuses on energy harvesting with all its antennas, i.e., $N_J = N_t + N_r$

---

[3]But we assume that the channel distribution of the eavesdropper is available. This assumption has been widely adopted in the literature for secrecy performance analysis [9, 17].

Fig. 2.8 A schematic diagram of the proposed accumulate-and-jam protocol

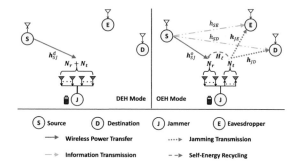

antennas, to receive RF signals. By doing so, J can maximize its energy acquisition by leaving no antennas idle. To enable the aforementioned functionality, J is also equipped with the following components[4]:

- N_J RF chains for energy harvesting and jamming transmission;
- N_J rectifiers for rectifying the received RF signals into direct currents (DC) [23, 24];
- a primary energy storage (PES), i.e. a chemical rechargeable battery with high energy density;
- a secondary energy storage (SES), i.e. a super-capacitor with high power density.

Specifically, N_r out of the N_J antennas are connected permanently to N_r rectifiers. The rest ones, i.e., $N_t = N_J - N_r$, are connected to the remaining N_t rectifiers in a non-permanent manner. For simplicity, we consider this antenna allocation as predetermined, and the potential antenna selection problem is beyond the scope of this work. The reason for employing the hybrid energy storage system at J is that a single energy storage cannot be charged and discharged at the same time, and therefore cannot support the FD operation. Briefly, PES is directly connected to the rectifiers and the RF chains. When the RF chains are idle, the harvested energy is delivered straight into PES. During transmission, PES uses its stored energy to power up the RF chains. Meanwhile, the harvested energy is temporarily saved in SES. Once the jamming transmission finishes, SES transfers all its stored energy to PES.

It is also important to clarify that the FD technique applied at J is for simultaneous energy reception and information transmission, which does not strictly fall into the category of conventional FD communications. With a slightly abused terminology, we refer such a jammer as a wireless-powered FD node.[5] In addition, for the comparison purpose, we also consider the usage of a wireless-powered HD jammer,

---

[4]Note that at the current stage of research, the optimal structure of an RF energy harvesting node is not completely known. The proposed circuit model in this work provides only one possible practical design.

[5]It is also worth noting that such unconventional usage of the term full-duplex is not uncommon in existing literature [8, 25, 26].

which performs either energy harvesting or jamming transmission, but not at the same time. More details are given in Sect. 2.2.5.

2.2.1.2 Channel Assumptions

In Fig. 2.8, fading channel coefficients of the links $S \rightarrow J$, $S \rightarrow D$, $S \rightarrow E$, $J \rightarrow D$, and $J \rightarrow E$ are denoted by $\mathbf{h}_{SJ}$, h_{SD}, h_{SE}, $\mathbf{h}_{JD}$, and $\mathbf{h}_{JE}$, respectively. We assume a quasi-static flat fading channel model, in which these fading channel coefficients remain constant within each transmission block of duration T,[6] and vary independently from one block to another. We apply different small-scale fading models to the channel $\mathbf{h}_{SJ}$ that performs energy harvesting and the other channels that perform information transmission. Specifically, since the up-to-date wireless energy harvesting techniques will only work within a relatively short distance, a line-of-sight (LoS) path is likely to present between S and J. Therefore, following [27, 28], we model $\mathbf{h}_{SJ}$ as a Rician fading channel. Through adjusting the Rician factor K, different channels can be modeled, ranging from a fully deterministic LoS channel (i.e. for short distance) to a weakly dominated LoS channel (i.e. for relatively large distance). On the other hand, for all other channels performing information transmission, we apply an independent and identically distributed (i.i.d.) Rayleigh fading to model the heavily scattered wireless communication.

In this chapter, we also make the following assumptions regarding the CSI: h_{SD} and $\mathbf{h}_{JD}$ are acquired respectively by S and J via channel estimation, but the estimated $\mathbf{h}_{JD}$ at J is imperfect. The CSI of the eavesdropper, i.e., h_{SE} and $\mathbf{h}_{JE}$, is known only to itself as a result of passive eavesdropping. The rationality and advantages of considering imperfect $\mathbf{h}_{JD}$ at J are two-fold: (1) As an energy-constrained node, the jammer has limited processing power and capability to perform accurate channel estimation, and (2) With imperfect $\mathbf{h}_{JD}$, we have extended the nulling jamming scheme to allow for interference leakage at the destination. We, therefore, are able to evaluate the impact of imperfect CSI on the system performance. Finally, channel reciprocity is assumed for all the wireless links in the considered system.

2.2.1.3 Protocol Description

In this section, we provide a detailed description of the proposed accumulate-and-jam (AnJ) protocol. Specifically, at the beginning of the kth transmission block ($k = 1, 2, \ldots$), J estimates its residual energy $\varepsilon[k]$ and compares it with a predefined threshold E_{th}. In the case of $\varepsilon[k] \geq E_{th}$ or $\varepsilon[k] < E_{th}$, J broadcasts a single bit (i.e.

---

[6]Without loss of generality, we normalize the block duration to one time unit, i.e., $T = 1$. As a consequence, the measures of energy and power become identical in this work and therefore can be used interchangeably.

1 or 0) to inform S and D whether it is capable of CJ, with bit 0 indicating that the energy at J is not sufficient for CJ, and bit 1 otherwise. If bit 0 is received from J, S feeds back bit 0 to J and D via a feedback channel to indicate that the current block will operate in DEH mode, and thus D keeps silent during this block. Otherwise (i.e. S and D receives bit 1 from J), S keeps listening, and D sends a pilot signal for S and J to perform channel estimation. With the assumed channel reciprocity, S can estimate h_{SD} which is then used to verify whether the instantaneous channel capacity C_{SD}, expressed as,

$$C_{SD} = \log_2\left(1 + \frac{P_S H_{SD}}{\sigma_D^2}\right) \qquad (2.2.33)$$

can support a secrecy rate R_s. In (2.2.33), P_S is the source transmitting power, $H_{SD} = |h_{SD}|^2$, and σ_D^2 is the additive white Gaussian noise (AWGN) at D. When $C_{SD} \geq R_s$, S feeds back bit 1 to D and J to indicate that the current block will operate in OEH mode. Otherwise (i.e. $C_{SD} < R_s$), S feeds back bit 0 to indicate that the DEH mode will be activated. The signaling message exchange among S, D, and J then terminates here. Let $\Phi[k] \in \{\Phi_d, \Phi_o\}$ indicate the operation mode (i.e., either DEH or OEH) for the kth transmission block, we have

$$\Phi[k] = \begin{cases} \Phi_o, & \text{if } \varepsilon[k] \geq E_{th} \text{ and } C_{SD} \geq R_s, \\ \Phi_d, & \text{otherwise.} \end{cases} \qquad (2.2.34)$$

The condition of $\varepsilon[k] > E_{th}$ is referred to as the "*energy condition*". It is used to prevent the intensity of the jamming signal from dropping down to the noise level at E, which wastes the acquired energy at J. The condition of $C_{SD} > R_s$ is referred to as the "*channel condition*". The system will suffer from secrecy outage if the channel condition is not met. The necessity of the channel condition will be shown clearly in Sect. 2.2.3 where the secrecy outage is defined.

In the following, we present the details of the signal processing occurred in each mode. In the DEH mode, with a fixed transmitting power P_S, S sends J an energy-bearing signal, which is randomly generated and contains no secret information. To maximize the acquired energy, J employs all the N_J antennas for receiving RF signals. By ignoring the negligible energy harvested from the receiver noise, the total amount of energy harvested at J during a DEH block is given by [11]

$$E_h^d = \eta P_S H_{SJ}^d, \qquad (2.2.35)$$

where $H_{SJ}^d = \left\|\mathbf{h}_{SJ}^d\right\|^2$, $\mathbf{h}_{SJ}^d \in \mathbb{C}^{N_J \times 1}$, and η denotes the energy conversion efficiency. The harvested energy is delivered straight into PES.

In the OEH mode, with the same transmitting power P_S, S sends an information-bearing signal x_S to D, with $\mathbb{E}\{|x_S|^2\} = 1$. J sends a jamming signal $\mathbf{x}_J$, which is deliberately designed for the purpose of producing a null at D and degrading the wiretap channel of E. It is clear that only when J is equipped with $N_t \geq 2$ antennas there are enough degrees of freedom to design $\mathbf{x}_J$. Specifically, the artificial noise generation method proposed in [5] is adopted, which requires instantaneous CSI $\mathbf{h}_{JD}$ for beam design.[7] Unfortunately, as an energy-constrained node, the processing capability of the jammer is limited. Therefore, there is a certain degree of mismatch between the estimated CSI $\hat{\mathbf{h}}_{JD}$ and the real CSI $\mathbf{h}_{JD}$, of which relation can be expressed as [32, 33]

$$\mathbf{h}_{JD} = \sqrt{\rho}\,\hat{\mathbf{h}}_{JD} + \sqrt{1 - \rho}\,\mathbf{h}_{err} \tag{2.2.36}$$

where $\mathbf{h}_{err}$ is the error noise vector with i.i.d. zero mean and variance σ_{err}^2. ρ, scaling from 0 to 1, is the correlation coefficient between $\hat{\mathbf{h}}_{JD}$ and $\mathbf{h}_{JD}$. A larger ρ means better CSI accuracy. If $\rho = 1$, J has full CSI $\mathbf{h}_{JD}$.

As mentioned earlier, the imperfect estimate $\hat{\mathbf{h}}_{JD}$ is used to design the jamming signal $\mathbf{x}_J$. Specifically, $\mathbf{x}_J = \mathbf{W}\mathbf{v}$, where $\mathbf{W}$ is a $N_t \times (N_t - 1)$ matrix constructed in the null-space of $\hat{\mathbf{h}}_{JD}$, and $\mathbf{v}$ is the artificial noise vector with $N_t - 1$ elements. Each element of $\mathbf{v}$ is assumed to be an i.i.d. complex Gaussian random variable with zero mean and normalized variance. Thus, the received signal at D and E are given by

$$y_D = \sqrt{P_S}h_{SD}x_S + \sqrt{(1-\rho)P_J}\mathbf{h}_{err}^\dagger\frac{\mathbf{W}\mathbf{v}}{\sqrt{N_t - 1}} + n_D \tag{2.2.37}$$

and

$$y_E = \sqrt{P_S}h_{SE}x_S + \sqrt{P_J}\mathbf{h}_{JE}^\dagger\frac{\mathbf{W}\mathbf{v}}{\sqrt{N_t - 1}} + n_E, \tag{2.2.38}$$

respectively, where P_J is the transmitting power of J (i.e., $0 < P_J < E_{th}$). n_D and n_E denote the AWGN with zero mean and variance σ_D^2 and σ_E^2, respectively. It can be seen from (2.2.37) that the jamming signal also leaks into D's receiver due to the estimation error. We show later at the numerical results in Sect. 2.2.6 the impact of ρ on the protocol performance.

Apart from information transmission and reception, wireless energy harvesting continues in OEH mode. Specifically, the received signal at J is given by

$$y_J = \sqrt{P_S}\mathbf{h}_{SJ}^o x_S + \sqrt{P_J}\mathbf{H}_\ell\frac{\mathbf{W}\mathbf{v}}{\sqrt{N_t - 1}} + \mathbf{n}_J \tag{2.2.39}$$

where $\mathbf{H}_\ell \in \mathbb{C}^{N_r \times N_t}$ denotes the loop channel at J, $\mathbf{n}_J$ is the $N_r \times 1$ AWGN vector satisfying $\mathbb{E}[\mathbf{n}_J\mathbf{n}_J^\dagger] = \mathbf{I}_{N_r}\sigma_J^2$, and σ_J^2 is the noise variance at each receiving antenna.

From (2.2.39), the energy-harvesting circuitry of J not only harvests energy from the signal that it overhears from S, but also recycles a portion of its own transmitted

---

[7]The jammer may estimate $\mathbf{h}_{JD}$ via channel training methods, e.g., [19, 29–31].

energy. However, to the best of our knowledge, the distribution of the loop channel $\mathbf{H}_\ell$ before applying any interference cancellation is still unknown in open literature. As a result, it is extremely difficult to derive the statistical functions for the recycled energy term. To make the ensuing mathematical analysis tractable and to attain meaningful results, we have to omit the recycled energy term in the following theoretical derivations. In this case, the total amount of harvested energy in OEH mode is expressed as

$$E_h^o \approx \eta P_S H_{SJ}^o \tag{2.2.40}$$

where $H_{SJ}^o := \left\| \mathbf{h}_{SJ}^o \right\|^2, \mathbf{h}_{SJ}^o \in \mathbb{C}^{N_r \times 1}$. Similar to (2.2.35), the harvested noise power is also ignored in (2.2.40). The harvested energy is first saved at SES and then delivered to PES once the jamming transmission finishes. It is noteworthy that due to the omission of recycled energy, strictly speaking, the ensuing theoretical analysis in this work draws a lower bound for the secrecy performance of the proposed AnJ protocol.

2.2.2 Hybrid Energy Storage State Analysis

The purpose of studying the jammer's energy storage is to find out the probability that the energy condition is met. Due to the FD operation mode, the energy status at J exhibits a complex charging and discharging behavior. We tackle this problem by first applying energy discretization at PES, then using a finite-state MC to model the state transition between discrete energy levels.

2.2.2.1 Energy Discretization

As we consider a practical energy storage with finite capacity, the analyses designed for infinite battery capacity [9] are not applicable to the current study. We, therefore, follow [12] and apply a discrete-level battery model to characterize the dynamic behaviors of PES and SES. Specifically, we discretize PES into $L + 1$ energy levels, with the i-th level expressed as $\varepsilon_i := iC_1/L, i \in \{0, 1, \ldots, L\}$, where C_1 represents the capacity of PES and is assumed to be greater than E_{th} (i.e., otherwise the jammer can never transmit).

Specifically, in DEH mode, the discretized energy saved in PES can be expressed as

$$\varepsilon_h^d := \varepsilon_{i_h^d}, \quad \text{where } i_h^d = \arg \max_{i \in \{0, 1, \ldots, L\}} \{\varepsilon_i : \varepsilon_i \leq E_h^d\}. \tag{2.2.41}$$

It is worth pointing out that if $E_h^d > C_1$, energy will overflow because the maximum amount of energy that can be saved in PES is C_1. Equation (2.2.41) implies $\varepsilon_h^d \leq \varepsilon_L = C_1$ by limiting $i \in \{0, 1, \ldots, L\}$. As for the OEH mode, since the acquired energy E_h^o is first saved in SES and then transferred to PES, considering energy overflow at SES, the amount of energy exported by SES is equal to $\min\{E_h^o, C_2\}$. After transferring along the circuitry from SES to PES, the amount of energy imported to PES can be expressed as

$$\tilde{E}_h^o = \eta' \times \min\{E_h^o, C_2\} \tag{2.2.42}$$

where η' is the energy transfer efficiency from SES to PES [21], and $\min\{x, y\}$ gives the smaller value between x and y. After discretization, the amount of energy eventually saved in PES is given by

$$\varepsilon_h^o := \varepsilon_{i_h^o}, \text{ where } i_h^o = \arg \max_{i \in \{0, 1, \ldots, L\}} \{\varepsilon_i : \varepsilon_i \leq \tilde{E}_h^o\}. \tag{2.2.43}$$

On the other hand, the required energy for jamming transmission E_{th} corresponds to a discrete energy level ε_t, which is defined as

$$\varepsilon_t := \varepsilon_{i_t}, \text{ where } i_t = \arg \min_{i \in \{0, 1, \ldots, L\}} \{\varepsilon_i : \varepsilon_i \geq E_{th}\}. \tag{2.2.44}$$

Note that E_{th} entails all energy consumption occurred at J, i.e., $E_{th} = P_J + P_c$, where P_c denotes the constant circuitry power [34, 35]. Furthermore, (2.2.44) can also be expressed as

$$\varepsilon_t = \left\lceil \frac{E_{th}}{C_1/L} \right\rceil \frac{C_1}{L} = \frac{\tau}{L} C_1, \tag{2.2.45}$$

where $\lceil \cdot \rceil$ stands for the ceiling function, and $\tau := \left\lceil \frac{E_{th}}{C_1/L} \right\rceil$ is defined for notation simplicity.

At the beginning of the $[k+1]$th transmission block, the residual energy $\varepsilon[k+1]$ is determined by the operation mode $\Phi[k]$ and the residual energy $\varepsilon[k]$ from the kth block. Therefore,

$$\varepsilon[k+1] = \begin{cases} \min\{\varepsilon[k] - \varepsilon_t + \varepsilon_h^o, C_1\} & \text{if } \Phi[k] = \Phi_o, \\ \min\{\varepsilon[k] + \varepsilon_h^d, C_1\} & \text{if } \Phi[k] = \Phi_d. \end{cases} \tag{2.2.46}$$

It is worth noting that the discrete energy model can tightly approximate its continuous counterpart when the number of the discretization level is sufficiently large [36] ($L \to \infty$ corresponds to a continuous energy storage model). The impact of L on the secrecy performance of the proposed protocol is presented in Sect. 2.2.6.

2.2.2.2   Markov Chain

With the energy discretization described above, we are able to model the transition of the PES energy states as a finite-state MC.[8] We define state S_i as the residual energy of PES $\varepsilon[k]$ being ε_i. The transition probability $p_{i,j}$ represents the probability of a transition from state S_i to state S_j. The transitions of the PES energy states have the following six cases:

1. *PES remains empty ($S_0 \rightarrow S_0$):* In this case, the energy condition cannot be met. Therefore, the DEH mode is activated. Provided that PES remains empty after recharging, it indicates that the harvested energy during this DEH block is discretized to zero, i.e., $\varepsilon_h^d = \varepsilon_0 = 0$. From the definition given in (2.2.35) and (2.2.41), the condition of $E_h^d = \eta P_S H_{SJ} < \varepsilon_1 = C_1/L$ must remain if the harvested energy is discretized to zero. The transition probability $p_{0,0}$ is thus described as

$$
\begin{aligned}
p_{0,0} &= \Pr\left\{\varepsilon_h^d = 0\right\} \\
&= \Pr\left\{E_h^d < \varepsilon_1\right\} \\
&= \Pr\left\{H_{SJ}^d < \frac{1}{\eta P_S L/C_1}\right\}
\end{aligned}
\tag{2.2.47}
$$

Since the channel between S and J is assumed to be Rician fading, the CDF of H_{SJ}^d is given by [37]

$$
F_{H_{SJ}^d}(x) = 1 - Q_{N_J}\left(\sqrt{2N_J K}, \sqrt{\frac{2(K+1)}{\Omega_{SJ}}x}\right),
\tag{2.2.48}
$$

where $Q_{N_J}(\cdot, \cdot)$ is the generalized (N_J-th order) Macum Q-function [38], and K is the rician factor. Combining (2.2.48) with (2.2.47), we have

$$
p_{0,0} = F_{H_{SJ}^d}\left(\frac{1}{\eta P_S L/C_1}\right)
\tag{2.2.49}
$$

2. *PES remains full ($S_L \rightarrow S_L$):* In this case, the energy condition is certainly met. The selection of the operation mode thus depends purely on the channel condition. If $C_{SD} \geq R_s$, OEH is invoked, where the consumed energy ε_t is no larger than the harvested energy ε_h^o. Otherwise (i.e. $C_{SD} < R_s$), DEH is invoked, and the harvested energy ε_h^d can be any arbitrary value as PES is full and cannot accept more energy. The corresponding transition probability is calculated as

---

[8]Note that it is not necessary to model the state transition of the SES since it is used only for temporary energy storage in OEH mode.

$$p_{L,L} = \Pr\{C_{SD} \geq R_s\} \Pr\{\varepsilon_h^o \geq \varepsilon_t\} + \Pr\{C_{SD} < R_s\} \qquad (2.2.50)$$

We first evaluate q_c, i.e. the probability of the channel condition being satisfied. After performing some simple manipulations, we have

$$q_c = \Pr\{C_{SD} \geq R_s\} = \Pr\left\{H_{SD} \geq \frac{2^{R_s} - 1}{P_S/\sigma_D^2}\right\} \qquad (2.2.51)$$

Since the channel h_{SD} is assumed to be Rayleigh fading, H_{SD} follows an exponential distribution with CDF

$$F_{H_{SD}}(x) = 1 - \exp\left(-\frac{x}{\Omega_{SD}}\right) \qquad (2.2.52)$$

Consequently, the probability for the channel condition is given by

$$q_c = 1 - F_{H_{SD}}\left(\frac{2^{R_s} - 1}{P_S/\sigma_D^2}\right) \qquad (2.2.53)$$

and

$$\Pr\{C_{SD} < R_s\} = 1 - q_c = F_{H_{SD}}\left(\frac{2^{R_s} - 1}{P_S/\sigma_D^2}\right) \qquad (2.2.54)$$

We now analyze the term $\Pr\{\varepsilon_h^o \geq \varepsilon_t\}$. From the definition given in (2.2.42), we have

$$
\begin{aligned}
\Pr\{\varepsilon_h^o \geq \varepsilon_t\} &= \Pr\{(\eta' E_h^o \geq \varepsilon_t) \cap (E_h^o < C_2)\} \\
&\quad + \Pr\{(\eta' C_2 \geq \varepsilon_t) \cap (E_h^o \geq C_2)\} \\
&= \begin{cases} \Pr\{\eta' E_h^o \geq \varepsilon_t\} & \text{if } C_2 \geq \frac{\tau}{\eta' L} C_1 \\[2mm] 0 & \text{otherwise} \end{cases}
\end{aligned} \qquad (2.2.55)
$$

With the definition of E_h^o given in (2.2.40), we can obtain

$$\Pr\{\eta' E_h^o \geq \varepsilon_t\} = \Pr\left\{H_{SJ}^o \geq \frac{\tau}{\eta\eta' P_S L/C_1}\right\} \qquad (2.2.56)$$

Similar to (2.2.48), the CDF of H_{SJ}^o is

$$F_{H_{SJ}^o}(x) = 1 - Q_{N_r}\left(\sqrt{2N_r K}, \sqrt{\frac{2(K+1)}{\Omega_{SJ}}x}\right) \qquad (2.2.57)$$

Consequently,

$$\Pr\{\eta' E_h^o \geq \varepsilon_t\} = 1 - F_{H_{SJ}^o} \left(\frac{\tau}{\eta\eta' P_S L/C_1} \right) \tag{2.2.58}$$

By combining (2.2.53), (2.2.54), (2.2.55), and (2.2.58), we can obtain the transition probability $p_{L,L}$ as

$$p_{L,L} = \begin{cases} q_c \left(1 - F_{H_{SJ}^o} \left(\frac{\tau}{\eta\eta' P_S L/C_1} \right)\right) & \text{if } C_2 \geq \frac{\tau}{\eta' L} C_1 \\ \\ F_{H_{SD}} \left(\frac{2^{R_s}-1}{P_S/\sigma_D^2} \right) & \text{otherwise} \end{cases} \tag{2.2.59}$$

3. *The non-empty and non-full PES remains unchanged ($S_i \rightarrow S_i : 0 < i < L$):* In this transition case, we need to first evaluate the energy condition. If the available energy is less than the required energy threshold, i.e., $\varepsilon_i < \varepsilon_t$, DEH mode is invoked. If $\varepsilon_i \geq \varepsilon_t$, then the energy condition is met. Next, we need to evaluate the channel condition. In the case that the channel condition is not satisfied, i.e., $C_{SD} < R_S$, again DEH mode is selected. Similar to the first transition probability (i.e., $S_0 \rightarrow S_0$), the unchangeable state transition in DEH mode indicates that the harvested energy is discretized to zero, i.e., $\varepsilon_h^d = 0$. On the other hand, if $\varepsilon_i \geq \varepsilon_t$ and $C_{SD} \geq R_S$ are both satisfied, OEH is activated. The unchangeable state transition in OEH mode indicates that the amount of harvested energy must equal the amount of the consumed energy, i.e., $\varepsilon_h^o = \varepsilon_t$. The transition probability is thus described as

$$\begin{aligned} p_{i,i} &= \Pr\{\varepsilon_i < \varepsilon_t\} \Pr\{\varepsilon_h^d = 0\} \\ &\quad + \Pr\{\varepsilon_i \geq \varepsilon_t\} \Pr\{C_{SD} < R_s\} \Pr\{\varepsilon_h^d = 0\} \\ &\quad + \Pr\{\varepsilon_i \geq \varepsilon_t\} \Pr\{C_{SD} \geq R_s\} \Pr\{\varepsilon_h^o = \varepsilon_t\} \\ \\ &= \begin{cases} \Pr\{\varepsilon_h^d = 0\} & \text{if } i < \tau \\ \\ \Pr\{C_{SD} < R_s\} \Pr\{\varepsilon_h^d = 0\} \\ \quad + \Pr\{C_{SD} \geq R_s\} \Pr\{\varepsilon_h^o = \varepsilon_t\} & \text{if } i \geq \tau \end{cases} \end{aligned} \tag{2.2.60}$$

From the definition given in (2.2.42), we have

$$\begin{aligned} \Pr\{\varepsilon_h^o = \varepsilon_t\} &= \Pr\{(0 \leq \eta' E_h^o - \varepsilon_t < \varepsilon_1) \cap (E_h^o < C_2)\} \\ &\quad + \Pr\{(0 \leq \eta' C_2 - \varepsilon_t < \varepsilon_1) \cap (E_h^o \geq C_2)\} \end{aligned}$$

$$= \begin{cases} 0 & \text{if } C_2 < \frac{\tau}{\eta' L} C_1 \\[2mm] \Pr\{\eta' E_h^o \geq \varepsilon_t\} & \text{if } \frac{\tau}{\eta' L} C_1 \leq C_2 < \frac{\tau+1}{\eta' L} C_1 \\[2mm] \Pr\{\varepsilon_t \leq \eta' E_h^o < \varepsilon_1 + \varepsilon_t\} & \text{if } C_2 > \frac{\tau+1}{\eta' L} C_1 \end{cases}$$

$$(2.2.61)$$

With some simple manipulations, after combining (2.2.47), (2.2.49), (2.2.53), (2.2.54), (2.2.58) and (2.2.61), the transition probability $p_{i,i}$ is given in (2.2.62).

$$p_{i,i} = \begin{cases} F_{H_{SJ}^d}\left(\frac{1}{\eta P_S L / C_1}\right) & \text{if } i < \tau \\[3mm] (1-q_c) F_{H_{SJ}^d}\left(\frac{1}{\eta P_S L / C_1}\right) & \text{if } i \geq \tau \ \& \ C_2 < \frac{\tau}{\eta' L} C_1 \\[3mm] (1-q_c) F_{H_{SJ}^d}\left(\frac{1}{\eta P_S L / C_1}\right) & \text{if } i \geq \tau \ \& \ \frac{\tau}{\eta' L} C_1 \leq C_2 < \frac{\tau+1}{\eta' L} C_1 \\ \quad + q_c \left(1 - F_{H_{SJ}^o}\left(\frac{\tau}{\eta\eta' P_S L / C_1}\right)\right) \\[3mm] (1-q_c) F_{H_{SJ}^d}\left(\frac{1}{\eta P_S L / C_1}\right) & \text{if } i \geq \tau \ \& \ C_2 \geq \frac{\tau+1}{\eta' L} C_1 \\ \quad + q_c \left(F_{H_{SJ}^o}\left(\frac{\tau+1}{\eta\eta' P_S L / C_1}\right) - F_{H_{SJ}^o}\left(\frac{\tau}{\eta\eta' P_S L / C_1}\right)\right) \end{cases}$$

$$(2.2.62)$$

4. *PES is fully charged ($S_i \rightarrow S_L : 0 \leq i < L$):* In this case, the harvested energy after discretization satisfies $\varepsilon_h^d \geq \varepsilon_L - \varepsilon_i$ in DEH, or $\varepsilon_h^o - \varepsilon_t \geq \varepsilon_L - \varepsilon_i$ in OEH. The transition probability $p_{i,L}$ is thus described as

$$\begin{aligned} p_{i,L} &= \Pr\{\varepsilon_i < \varepsilon_t\} \Pr\{\varepsilon_h^d \geq \varepsilon_L - \varepsilon_i\} \\ &\quad + \Pr\{\varepsilon_i \geq \varepsilon_t\} \Pr\{C_{SD} < R_s\} \Pr\{\varepsilon_h^d \geq \varepsilon_L - \varepsilon_i\} \\ &\quad + \Pr\{\varepsilon_i \geq \varepsilon_t\} \Pr\{C_{SD} \geq R_s\} \Pr\{\varepsilon_h^o - \varepsilon_t \geq \varepsilon_L - \varepsilon_i\} \\ &= \begin{cases} \Pr\{\varepsilon_h^d \geq \varepsilon_L - \varepsilon_i\} & \text{if } i < \tau \\[3mm] (1-q_c) \Pr\{\varepsilon_h^d \geq \varepsilon_L - \varepsilon_i\} & \text{if } i \geq \tau \\ \quad + q_c \Pr\{\varepsilon_h^o - \varepsilon_t \geq \varepsilon_L - \varepsilon_i\} \end{cases} \end{aligned}$$

$$(2.2.63)$$

Particularly, we have

$$\Pr\{\varepsilon_h^d \geq \varepsilon_L - \varepsilon_i\} = \Pr\{E_h^d \geq C_1 - \varepsilon_i\} \tag{2.2.64}$$

and

$$\Pr\{\varepsilon_h^o - \varepsilon_t \geq \varepsilon_L - \varepsilon_i\}$$
$$= \Pr\{(\eta' E_h^o - \varepsilon_t \geq C_1 - \varepsilon_i) \cap (E_h^o < C_2)\}$$
$$+ \Pr\{(\eta' C_2 - \varepsilon_t \geq C_1 - \varepsilon_i) \cap (E_h^o \geq C_2)\}$$
$$= \begin{cases} 0 & \text{if } C_2 < \frac{L-i+\tau}{\eta' L} C_1 \\ \Pr\{\eta' E_h^o \geq C_1 - \varepsilon_i + \varepsilon_t\} & \text{if } C_2 \geq \frac{L-i+\tau}{\eta' L} C_1 \end{cases} \quad (2.2.65)$$

Consequently, the transition probability in this case is given by (2.2.66).

$$p_{i,L} = \begin{cases} 1 - F_{H_{SJ}^d}\left(\frac{L-i}{\eta P_S L / C_1}\right) & \text{if } i < \tau \\\\ (1 - q_c)\left(1 - F_{H_{SJ}^d}\left(\frac{L-i}{\eta P_S L / C_1}\right)\right) & \text{if } i \geq \tau \ \& \ C_2 < \frac{L-i+\tau}{\eta' L} C_1 \\\\ (1 - q_c)\left(1 - F_{H_{SJ}^d}\left(\frac{L-i}{\eta P_S L / C_1}\right)\right) & \text{if } i \geq \tau \ \& \ C_2 \geq \frac{L-i+\tau}{\eta' L} C_1 \\ \quad + q_c\left(1 - F_{H_{SJ}^o}\left(\frac{L-i+\tau}{\eta \eta' P_S L / C_1}\right)\right) \end{cases}$$

$$(2.2.66)$$

5. *PES is partially charged ($S_i \to S_j : 0 \leq i < j < L$):* This transition case can happen either in DEH mode with $\varepsilon_h^d = \varepsilon_j - \varepsilon_i$, or in OEH mode with $\varepsilon_h^o - \varepsilon_t = \varepsilon_j - \varepsilon_i$. The transition probability $p_{i,j}$ is therefore calculated as

$$p_{i,j} = \Pr\{\varepsilon_i < \varepsilon_t\} \Pr\{\varepsilon_h^d = \varepsilon_j - \varepsilon_i\}$$
$$+ \Pr\{\varepsilon_i \geq \varepsilon_t\} \Pr\{C_{SD} < R_s\} \Pr\{\varepsilon_h^d = \varepsilon_j - \varepsilon_i\}$$
$$+ \Pr\{\varepsilon_i \geq \varepsilon_t\} \Pr\{C_{SD} \geq R_s\} \Pr\{\varepsilon_h^o - \varepsilon_t = \varepsilon_j - \varepsilon_i\}$$
$$= \begin{cases} \Pr\{\varepsilon_h^d = \varepsilon_j - \varepsilon_i\} & \text{if } i < \tau \\\\ (1 - q_c) \Pr\{\varepsilon_h^d = \varepsilon_j - \varepsilon_i\} + q_c \Pr\{\varepsilon_h^o - \varepsilon_t = \varepsilon_j - \varepsilon_i\} & \text{if } i \geq \tau \end{cases}$$
$$(2.2.67)$$

Specifically, we have

$$\Pr\{\varepsilon_h^d = \varepsilon_j - \varepsilon_i\} = \Pr\{\varepsilon_j - \varepsilon_i \leq \varepsilon_h^d < \varepsilon_{j+1} - \varepsilon_i\} \quad (2.2.68)$$

and

$$\Pr\{\varepsilon_h^o - \varepsilon_t = \varepsilon_j - \varepsilon_i\}$$

$$= \Pr\{(\varepsilon_j - \varepsilon_i \leq \eta' E_h^o - \varepsilon_t < \varepsilon_{j+1} - \varepsilon_i) \cap (E_h^o < C_2)\}$$

$$\quad + \Pr\{(\varepsilon_j - \varepsilon_i \leq \eta' C_2 - \varepsilon_t < \varepsilon_{j+1} - \varepsilon_i) \cap (E_h^o \geq C_2)\}$$

$$= \begin{cases} 0 & \text{if } C_2 < \frac{j-i+\tau}{\eta' L} C_1 \\[2mm] \Pr\{\eta' E_h^o \geq \varepsilon_j - \varepsilon_i + \varepsilon_t\} & \text{if } \frac{j-i+\tau}{\eta' L} C_1 \leq C_2 < \frac{j-i+\tau+1}{\eta' L} C_1 \\[2mm] \Pr\{\varepsilon_j - \varepsilon_i + \varepsilon_t \leq \eta' E_h^o < \varepsilon_{j+1} - \varepsilon_i + \varepsilon_t\} & \text{if } C_2 > \frac{j-i+\tau+1}{\eta' L} C_1 \end{cases}$$

$$(2.2.69)$$

Therefore, we can obtain the transition probability in this case given by (2.2.70).

$$p_{i,j} = \begin{cases} F_{H_{SJ}^d}\left(\frac{j-i+1}{\eta P_S L/C_1}\right) - F_{H_{SJ}^d}\left(\frac{j-i}{\eta P_S L/C_1}\right) & \text{if } i < \tau \\[3mm] (1-q_c)\left(F_{H_{SJ}^d}\left(\frac{j-i+1}{\eta P_S L/C_1}\right) - F_{H_{SJ}^d}\left(\frac{j-i}{\eta P_S L/C_1}\right)\right) & \text{if } i \geq \tau \ \& \ C_2 < \frac{j-i+\tau}{\eta' L} C_1 \\[3mm] (1-q_c)\left(F_{H_{SJ}^d}\left(\frac{j-i+1}{\eta P_S L/C_1}\right) - F_{H_{SJ}^d}\left(\frac{j-i}{\eta P_S L/C_1}\right)\right) & \text{if } i \geq \tau \\ \quad + q_c\left(1 - F_{H_{SJ}^o}\left(\frac{j-i+\tau}{\eta\eta' P_S L/C_1}\right)\right) & \quad \& \ \frac{j-i+\tau}{\eta' L} C_1 \leq C_2 < \frac{j-i+\tau+1}{\eta' L} C_1 \\[3mm] (1-q_c)\left(F_{H_{SJ}^d}\left(\frac{j-i+1}{\eta P_S L/C_1}\right) - F_{H_{SJ}^d}\left(\frac{j-i}{\eta P_S L/C_1}\right)\right) & \text{if } i \geq \tau \ \& \ C_2 > \frac{j-i+\tau+1}{\eta' L} C_1 \\ \quad + q_c\left(F_{H_{SJ}^o}\left(\frac{j-i+\tau+1}{\eta\eta' P_S L/C_1}\right) - F_{H_{SJ}^o}\left(\frac{j-i+\tau}{\eta\eta' P_S L/C_1}\right)\right) & \end{cases}$$

$$(2.2.70)$$

6. *PES is discharged ($S_j \rightarrow S_i : 0 \leq i < j \leq L$):* Since the stored energy is reduced during this transition case, the OEH operation mode must have been activated. The amount of reduced energy, i.e., $\varepsilon_j - \varepsilon_i$, equals the difference between the consumed energy ε_t and the discretized acquired energy ε_h^o. The transition probability is expressed as

$$p_{j,i} = \Pr\{\varepsilon_j \geq \varepsilon_t\} \Pr\{C_{SD} \geq R_s\} \Pr\{\varepsilon_t - \varepsilon_h^o = \varepsilon_j - \varepsilon_i\}$$

$$= \begin{cases} 0 & \text{if } j < \tau \\ q_c \Pr\{\varepsilon_t - \varepsilon_h^o = \varepsilon_j - \varepsilon_i\} & \text{if } j \geq \tau \end{cases} \qquad (2.2.71)$$

And,

$$
\begin{aligned}
&\Pr\{\varepsilon_t - \varepsilon_h^o = \varepsilon_j - \varepsilon_i\} \\
&\quad = \Pr\{(\varepsilon_j - \varepsilon_{i+1} < \varepsilon_t - \eta' E_h^o \le \varepsilon_j - \varepsilon_i) \cap (E_h^o < C_2)\} \\
&\qquad + \Pr\{(\varepsilon_j - \varepsilon_{i+1} < \varepsilon_t - \eta' C_2 \le \varepsilon_j - \varepsilon_i) \cap (E_h^o \ge C_2)\} \\
&\quad = \begin{cases}
0 & \text{if } C_2 < \frac{\tau-j+i}{\eta'L}C_1 \\
\Pr\{\eta' E_h^o \ge \varepsilon_t - \varepsilon_j + \varepsilon_i\} & \text{if } \frac{\tau-j+i}{\eta'L}C_1 \le C_2 < \frac{\tau-j+i+1}{\eta'L}C_1 \\
\Pr\{\varepsilon_t - \varepsilon_j + \varepsilon_i \le \eta' E_h^o < \varepsilon_t - \varepsilon_i + \varepsilon_{i+1}\} & \text{if } C_2 \ge \frac{\tau-j+i+1}{\eta'L}C_1
\end{cases}
\end{aligned}
\tag{2.2.72}
$$

As a result, the transition probability in this case is given by (2.2.73).

$$
p_{j,i} = \begin{cases}
0 & \text{if } j < \tau \text{ or } C_2 < \frac{\tau-j+i}{\eta'L}C_1 \\
q_c\left(1 - F_{H_{SJ}^o}\left(\frac{\tau-j+i}{\eta\eta' P_S L/C_1}\right)\right) & \text{if } j \ge \tau \\
& \quad \& \frac{\tau-j+i}{\eta'L}C_1 \le C_2 < \frac{\tau-j+i+1}{\eta'L}C_1 \\
q_c\left(F_{H_{SJ}^o}\left(\frac{\tau-j+i+1}{\eta\eta' P_S L/C_1}\right) - F_{H_{SJ}^o}\left(\frac{\tau-j+i}{\eta\eta' P_S L/C_1}\right)\right) & \text{if } j \ge \tau \ \& \ C_2 \ge \frac{\tau-j+i+1}{\eta'L}C_1
\end{cases}
\tag{2.2.73}
$$

We then examine the stationary distribution $\boldsymbol{\xi}_{FD}$ of the PES energy status, where $\xi_{FD,i}, i \in \{0, 1, \ldots, L\}$ represents the probability of the residual energy at PES being ε_i. We first define $\mathbf{M}_{FD} := (p_{i,j})$ to denote the $(L+1) \times (L+1)$ state transition matrix of the MC. By using the similar methods in [39], we can easily verify that the MC transition matrix $\mathbf{M}_{FD}$ is irreducible and row stochastic. Therefore, there must exist a unique stationary distribution $\boldsymbol{\xi}_{FD}$ that satisfies the following equation

$$
\boldsymbol{\xi}_{FD} = (\xi_{FD,0}, \xi_{FD,1}, \ldots, \xi_{FD,L})^T = (\mathbf{M}_{FD})^T \boldsymbol{\xi}_{FD}
\tag{2.2.74}
$$

By solving (2.2.74), $\boldsymbol{\xi}_{FD}$ can be derived as

$$
\boldsymbol{\xi}_{FD} = \left((\mathbf{M}_{FD})^T - \mathbf{I} + \mathbf{B}\right)^{-1} \mathbf{b},
\tag{2.2.75}
$$

where $B_{i,j} = 1, \forall i, j$ and $\mathbf{b} = (1, 1, \ldots, 1)^T$.

We are now ready to derive the probability that the available energy at J mets the energy condition. With the stationary distribution $\boldsymbol{\xi}_{FD}$, we can obtain

$$
\Pr\{\varepsilon[k] \ge E_{th}\} = \sum_{i=\tau}^{L} \xi_{FD,i}
\tag{2.2.76}
$$

2.2.3 Performance Evaluation

In this section, we characterize the secrecy performance of the proposed AnJ protocol in terms of the secrecy outage probability and the existence of non-zero secrecy capacity. These two probabilistic metrics are widely used in measuring secrecy performance when the eavesdropper's instantaneous CSI is absent.

2.2.3.1 Preliminaries

The secrecy capacity C_s is defined as the rate difference between the maximum achievable transmission rate of the main channel and that of the wiretap channel [3]. Mathematically speaking,

$$C_s = \begin{cases} C_M - C_W & \text{if } \gamma_D > \gamma_E \\ 0 & \text{if } \gamma_D \leq \gamma_E \end{cases} \tag{2.2.77}$$

where $C_M = \log_2 (1 + \gamma_D)$ is the capacity of the main channel between S and D, and $C_W = \log_2 (1 + \gamma_E)$ is the capacity of the wiretap channel between S and E. In (2.2.77), γ_D and γ_E denote the instantaneous SNRs at D and E, respectively.

Specifically, the secrecy outage probability, i.e., P_{so}^{AnJ}, is defined as the probability that the secrecy capacity C_s is less than a target secrecy rate R_s[9] [4]. Mathematically speaking,

$$P_{so}^{\text{AnJ}} = \Pr\{C_s < R_s\} \tag{2.2.78}$$

The existence of non-zero secrecy capacity, i.e., P_{nzsc}^{AnJ}, is defined as the probability that the secrecy capacity is greater than zero, i.e.,

$$P_{nzsc}^{\text{AnJ}} = \Pr\{C_s > 0\} \tag{2.2.79}$$

2.2.3.2 Secrecy Outage Probability

We first derive the signal-to-interference-plus-noise ratio (SINR) and its corresponding PDF and CDF at D and E, respectively. From (2.2.37), the SINR at D is given by

$$\gamma_D = \frac{P_S H_{SD}}{(1 - \rho) P_J \sigma_{err}^2 / (N_t - 1) + \sigma_D^2} \tag{2.2.80}$$

Since h_{SD} is Rayleigh fading channel, γ_D follows an exponential distribution.

---

[9]From (2.2.77) and (2.2.78), it is clear that secrecy outage must happen if the channel capacity of link S → D is less than R_s, which motivates our energy condition presented in Sect. 2.2.1.3.

From (2.2.38), the SINR at E is given by

$$\gamma_E = \frac{P_S |h_{SE}|^2}{P_J ||(\mathbf{h}_{JE})^\dagger \mathbf{W}||^2 / (N_t - 1) + \sigma_E^2} \tag{2.2.81}$$

The PDF of γ_E depends on $|h_{SE}|^2$ and $||(\mathbf{h}_{JE})^\dagger \mathbf{W}||^2$. To proceed, we first define $X := P_S |h_{SE}|^2$. Recall that h_{SE} follows a complex normal distribution, and let Ω_{SE} denote its variance, we thus have the PDF of X as

$$f_X(x) = \frac{1}{P_S \Omega_{SE}} \exp\left(-\frac{x}{P_S \Omega_{SE}}\right) \tag{2.2.82}$$

We also define $Y := \frac{P_J ||(\mathbf{h}_{JE})^\dagger \mathbf{W}||^2}{N_t - 1}$. Since $||(\mathbf{h}_{JE})^\dagger||^2$ is a sum of i.i.d. exponential distributed random variables, and $\mathbf{W}$ is a unitary matrix, $||(\mathbf{h}_{JE})^\dagger \mathbf{W}||^2$ is also a sum of i.i.d. exponential distributed random variables [6]. Therefore, Y follows a Gamma distribution $\mathcal{G}(N_t - 1, P_J \Omega_{JE}/(N_t - 1))$ with the PDF given by

$$f_Y(y) = \frac{y^{N_t - 2} e^{-\frac{N_t - 1}{P_J \Omega_{JE}} y}}{\Gamma(N_t - 1) \left(\frac{P_J \Omega_{JE}}{N_t - 1}\right)^{N_t - 1}} \tag{2.2.83}$$

According to (2.2.81), the expression $\gamma_E = \frac{X}{Y + \sigma_E^2}$ then holds. Therefore, we can obtain the CDF of γ_E as

$$
\begin{aligned}
F_{\gamma_E}(z) = \Pr\{\gamma_E < z\} &= \Pr\left\{\frac{X}{Y + \sigma_E^2} < z\right\} \\
&= \int_0^\infty \int_0^{zy + z\sigma_E^2} f_X(x) f_Y(y) \, dx \, dy \\
&= 1 - e^{-\frac{z\sigma_E^2}{P_S \Omega_{SE}}} \left(\frac{N_t - 1}{\varphi z + N_t - 1}\right)^{N_t - 1}
\end{aligned}
\tag{2.2.84}
$$

where $\varphi := \frac{P_J \Omega_{JE}}{P_S \Omega_{SE}}$, and the integral is obtained from [38, Eq. (3.326.2)]. Correspondingly, the PDF of γ_E is obtained as

$$f_{\gamma_E}(z) = \frac{\sigma_E^2 \, e^{-\frac{z\sigma_E^2}{P_S \Omega_{SE}}}}{P_S \Omega_{SE}} \left(\frac{N_t - 1}{\varphi z + N_t - 1}\right)^{N_t - 1} + \varphi e^{-\frac{z\sigma_E^2}{P_S \Omega_{SE}}} \left(\frac{N_t - 1}{\varphi z + N_t - 1}\right)^{N_t} \tag{2.2.85}$$

Proposition 2.3 *The exact secrecy outage probability for the proposed AnJ protocol is derived as*

$$P_{so}^{\text{AnJ}} = \begin{cases} 1 - \varphi^{-1} \exp\left(-\frac{2^{R_s}-1}{\kappa_1 \Omega_{SD}}\right) \sum_{i=\tau}^{L} \xi_{FD,i} & \text{if } N_t = 2 \\ \quad \times \left(\frac{\sigma_E^2}{P_S \Omega_{SE}} \Psi_1(1, \mu_1, \beta_1) + \Psi_1(2, \mu_1, \beta_1)\right) & \\ \\ 1 - \beta_1^{N_t-1} \exp\left(-\frac{2^{R_s}-1}{\kappa_1 \Omega_{SD}}\right) \sum_{i=\tau}^{L} \xi_{FD,i} & \text{if } N_t \geq 3 \\ \quad \times \left(\frac{\sigma_E^2}{P_S \Omega_{SE}} \Psi_2(N_t-1, \mu_1, \beta_1) + (N_t-1)\Psi_2(N_t, \mu_1, \beta_1)\right) & \end{cases}$$

$$(2.2.86)$$

where

$$\beta_1 := \frac{N_t - 1}{\varphi}. \tag{2.2.87a}$$

$$\mu_1 := \frac{2^{R_s}}{\kappa_1 \Omega_{SD}} + \frac{\sigma_E^2}{P_S \Omega_{SE}}. \tag{2.2.87b}$$

$$\kappa_1 := \frac{P_S}{(1-\rho)P_J \sigma_{err}^2/(N_t-1) + \sigma_D^2}. \tag{2.2.87c}$$

and

$$\Psi_1(n, \mu, \beta) := (n-1)\beta^{-1} - (-\mu)^{n-1} e^{\beta\mu} \text{Ei}(-\beta\mu). \tag{2.2.88a}$$

$$\Psi_2(n, \mu, \beta) := \frac{1}{(n-1)!} \sum_{k=1}^{n-1} (k-1)!(-\mu)^{n-k-1}\beta^{-k} \tag{2.2.88b}$$

$$- \frac{(-\mu)^{n-1}}{(n-1)!} e^{\beta\mu} \text{Ei}(-\beta\mu). \tag{2.2.88c}$$

Proof See Appendix A. ∎

Corollary 2.2 *The probability of non-zero secrecy capacity is given by*

$$P_{nzsc}^{\text{AnJ}} = \begin{cases} \varphi^{-1} e^{-\beta_2 \mu_2} \sum_{i=\tau}^{L} \xi_{FD,i} & \text{if } N_t = 2 \\ \quad \times \left(\frac{\sigma_E^2}{P_S \Omega_{SE}} \Psi_1(1, \mu_2, \beta_1+\beta_2) + \Psi_1(2, \mu_2, \beta_1+\beta_2)\right) & \\ \quad + \exp\left(-\frac{2^{R_s}-1}{\kappa_2 \Omega_{SD}}\right) F_{\gamma_E}(\beta_2) \sum_{i=\tau}^{L} \xi_{FD,i} & \\ \\ \beta_1^{N_t-1} e^{-\beta_2 \mu_2} \sum_{i=\tau}^{L} \xi_{FD,i} & \text{if } N_t \geq 3 \\ \quad \times \left(\frac{\sigma_E^2}{P_S \Omega_{SE}} \Psi_2(N_t-1, \mu_2, \beta_1+\beta_2) + (N_t-1)\Psi_2(N_t, \mu_2, \beta_1+\beta_2)\right) & \\ \quad + \exp\left(-\frac{2^{R_s}-1}{\kappa_2 \Omega_{SD}}\right) F_{\gamma_E}(\beta_2) \sum_{i=\tau}^{L} \xi_{FD,i} & \end{cases}$$

$$(2.2.89)$$

where

$$\kappa_2 := P_S/\sigma_D^2. \tag{2.2.90a}$$

$$\beta_2 := \frac{(2^{R_s} - 1)\kappa_1}{\kappa_2}. \tag{2.2.90b}$$

$$\mu_2 := \frac{1}{\kappa_1 \Omega_{SD}} + \frac{\sigma_E^2}{P_S \Omega_{SE}}. \tag{2.2.90c}$$

Proof See Appendix A.   ∎

Remark 2.4 For a given source transmit power P_S, the probability that the accumulated energy at J is sufficient for jamming decreases with the increase of the jamming power P_J. Specifically, the threshold τ for the last summation term in (2.2.86) increases with P_J. Accordingly, the probability summation $\sum_{i=\tau}^{L} \xi_{FD,i}$ decreases with the increase of P_J, which has a negative effect on the secrecy outage. On the other hand, according to (2.2.80) and (2.2.81), increasing P_J will reduce the SINR at both D and E, but to a lesser extent for γ_D compared to γ_E as the null space jamming is designed. Consequently, larger P_J can lead to larger instantaneous secrecy capacity C_s, which has a positive effect on the secrecy outage. To make a long story short, a higher P_J is associated with lower jamming frequency but higher interference strength. Therefore, we deduce that there would be an optimal P_J^* that minimizes P_{so}^{AnJ}. This will be verified by numerical results in Sect. 2.2.6. Unfortunately, due to the complexity of the considered MC model, it is difficult to find a general closed-form solution for P_J^*. Nevertheless, for a given network setup, we can readily obtain P_J^* by performing a one-dimensional exhaustive search over the finite range of the discretized energy levels.

2.2.4 Continuous Energy Storage Model with Infinite Capacity

In the proposed AnJ protocol, we employ PES and SES with finite storage capacity at the wireless powered jammer. It is obvious that the system performance can be improved via increasing the capacity of PES and SES: A larger capacity can reduce the energy loss caused by energy overflow, thus the jammer can accumulate more energy for supporting jamming transmission. On the other hand, one can infer that the rate of the performance improvement actually decreases as the energy storage capacity increases, because energy overflow occurs more rarely as the capacity keeps increasing. Considering the device cost and size, a question then comes up: "For a given network setup, how much energy storage capacity and the corresponding discretization level are considered as adequate?" To answer this question, in this subsection, we analyze the upper bound of the system performance with infinite energy storage capacity, i.e., $C_1 \to \infty, C_2 \to \infty$.

To investigate the long-term behavior of the infinite energy storage, we need to compare the energy consumption E_{th} with $\mathbb{E}[\tilde{E}_h^o]$, which is the average amount of energy acquired by J in OEH mode. Specifically, $\mathbb{E}[\tilde{E}_h^o] > E_{th}$ means that, on average, the harvested energy merely in OEH mode can fully meet the required energy consumption at the jammer. In this case, the energy stored in PES steadily accumulates towards infinity during the communication process, which makes the jammer always meet the energy condition. On the other hand, when $\mathbb{E}[\tilde{E}_h^o] < E_{th}$, the harvested energy in OEH mode is, on average, less than the consumed energy. As a result, the energy level at PES exhibits a sawtooth wave and can never accumulate towards infinity. In this case, the total amount of harvested energy must equal the total amount of energy consumption in the long run. Mathematically, with q_c being the probability of meeting the channel condition, and q_b indicating the probability of activating the energy condition, we have

$$q_c q_b \mathbb{E}[\tilde{E}_h^o] + (1 - q_c q_b) \mathbb{E}[E_h^d] = q_c q_b E_{th} \tag{2.2.91}$$

Therefore,

$$q_b = \frac{\mathbb{E}[E_n^d]}{q_c \left(E_{th} + \mathbb{E}[E_n^d] - \mathbb{E}[\tilde{E}_h^o] \right)} \tag{2.2.92}$$

With the CDF of H_{SJ}^d in (2.2.48), we can calculate $\mathbb{E}[E_n^d]$ as

$$\mathbb{E}[E_n^d] = \eta P_S \mathbb{E}[H_{SJ}^d] = \eta P_S \int_0^\infty x F_{H_{SJ}^d}'(x) \, \mathrm{d}x = \eta P_S N_J \Omega_{SJ} \tag{2.2.93}$$

Similarly, with the CDF of H_{SJ}^o in (2.2.57), we have

$$\mathbb{E}[\tilde{E}_h^o] = \eta \eta' P_S \mathbb{E}[H_{SJ}^o] = \eta \eta' P_S N_r \Omega_{SJ} \tag{2.2.94}$$

When combining (2.2.93) and (2.2.94) with (2.2.92), we can obtain q_b as

$$q_b = \frac{\eta P_S N_J \Omega_{SJ}}{\exp\left(-\frac{(2^{R_s}-1)\sigma_D^2}{P_S \Omega_{SD}}\right) (E_{th} + \eta P_S N_J \Omega_{SJ} - \eta \eta' P_S N_r \Omega_{SJ})} \tag{2.2.95}$$

Corollary 2.3 *The closed-form expression of the secrecy outage probability for a cooperative jammer with an infinite capacity energy storage can be obtained by replacing $\sum_{i=\tau}^L \xi_{FD,i}$ in (2.2.86) with q_b.*

2.2.5 Cooperative Jamming by a Wireless-Powered Half-Duplex Jammer

In this subsection, we consider an alternative cooperative jamming protocol with a wireless-powered HD jammer J' to provide a benchmark for evaluating the performance of the proposed AnJ protocol.

In order to compare our proposed FD jammer and this HD jammer in a fair manner, we assume that J' is equipped with the same number of antennas and rectifiers as J, i.e., N_J RF antennas and N_J rectifiers. All antennas and rectifiers are connected in a non-permanent manner. Due to the HD mode, J' requires only one energy storage. We let the capacity of the energy storage at J' be C_1, same as that of the PES at J. It is noteworthy that the analytical approach in [9] for a HD jammer is not applicable to J', because our energy storage has a finite capacity, while the battery capacity in [9] was infinite. Similar to J, J' also operates in two modes, i.e., the energy harvesting (EH) mode and the cooperative jamming (CJ) mode. In EH mode, J' performs exactly like J and harvests the same amount of energy E_h^d. All acquired energy is saved in its single energy storage. In CJ mode, on the contrary, J' uses all N_J antennas to transmit jamming signals, and therefore, does not acquire any energy. Assuming that imperfect channel estimation also occurs at J', the corresponding SINR at D and E are given by

$$\gamma_D' = \frac{P_S H_{SD}}{(1-\rho)P_J \sigma_{err}^2/(N_J - 1) + \sigma_D^2} \tag{2.2.96}$$

and

$$\gamma_E' = \frac{P_S |h_{SE}|^2}{P_J ||(\mathbf{h}_{JE})^\dagger \mathbf{W}||^2/(N_J - 1) + \sigma_E^2} \tag{2.2.97}$$

Note that (2.2.96) and (2.2.97) differ from (2.2.80) and (2.2.81) only in the number of transmitting antennas, i.e., N_r is replaced with N_J.

Remark 2.5 Owing to the FD operation mode, it is evident that J harvests more energy than J'. Yet, how much additional energy that J can acquire mainly depends on how the antennas at J are assigned for energy harvesting and jamming transmission. The impact of antenna allocation on secrecy performance of the proposed AnJ protocol will be shown in numerical results.

2.2.5.1 Markov Chain for HD Jammer

We discretize the battery at J' in the way as described in Sect. 2.2.2.1 (i.e., the same discretization as J). Different from the MC presented in Sect. 2.2.2.2, all state transitions at J' without a decrease in energy levels refer to the EH mode, because no energy is harvested by J' in the CJ mode. The state transition probability of J', $p_{i,j}'$, is characterized in the following six cases. Due to space scarcity, we have skipped the full details.

1. *The battery remains empty* ($S_0 \rightarrow S_0$):

$$p'_{0,0} = \Pr\left\{\varepsilon_h^d = 0\right\} = F_{H_{SJ}^d}\left(\frac{1}{\eta P_S L/C_1}\right) \tag{2.2.98}$$

2. *The battery remains full* ($S_L \rightarrow S_L$):

$$p'_{L,L} = \Pr\{C_{SD} < R_s\} = 1 - q_c \tag{2.2.99}$$

3. *The non-empty and non-full PES remains unchangeable* ($S_i \rightarrow S_i : 0 < i < L$):

$$
\begin{aligned}
p'_{i,i} &= \Pr\{\varepsilon_i < \varepsilon_t\}\Pr\{\varepsilon_h^d = 0\} \\
&\quad + \Pr\{\varepsilon_i \geq \varepsilon_t\}\Pr\{C_{SD} < R_s\}\Pr\{\varepsilon_h^d = 0\} \\
&= \begin{cases} F_{H_{SJ}^d}\left(\frac{1}{\eta P_S L/C_1}\right) & \text{if } i < \tau \\ (1 - q_c)F_{H_{SJ}^d}\left(\frac{1}{\eta P_S L/C_1}\right) & \text{if } i \geq \tau \end{cases}
\end{aligned} \tag{2.2.100}
$$

4. *PES is partially charged* ($S_i \rightarrow S_j : 0 \leq i < j < L$):

$$
\begin{aligned}
p'_{i,j} &= \Pr\{\varepsilon_i < \varepsilon_t\}\Pr\{\varepsilon_h^d = \varepsilon_j - \varepsilon_i\} \\
&\quad + \Pr\{\varepsilon_i \geq \varepsilon_t\}\Pr\{C_{SD} < R_s\}\Pr\{\varepsilon_h^d = \varepsilon_j - \varepsilon_i\} \\
&= \begin{cases} F_{H_{SJ}^d}\left(\frac{j-i+1}{\eta P_S L/C_1}\right) - F_{H_{SJ}^d}\left(\frac{j-i}{\eta P_S L/C_1}\right) & \text{if } i < \tau \\ \left(F_{H_{SJ}^d}\left(\frac{j-i+1}{\eta P_S L/C_1}\right) - F_{H_{SJ}^d}\left(\frac{j-i}{\eta P_S L/C_1}\right)\right)(1 - q_c) & \text{if } i \geq \tau \end{cases}
\end{aligned}
$$
$$\tag{2.2.101}$$

5. *The battery is fully charged* ($S_i \rightarrow S_L : 0 \leq i < L$):

$$
\begin{aligned}
p'_{i,L} &= \Pr\{\varepsilon_i < \varepsilon_t\}\Pr\{\varepsilon_h^d \geq \varepsilon_L - \varepsilon_i\} \\
&\quad + \Pr\{\varepsilon_i \geq \varepsilon_t\}\Pr\{C_{SD} < R_s\}\Pr\{\varepsilon_h^d \geq \varepsilon_L - \varepsilon_i\} \\
&= \begin{cases} 1 - F_{H_{SJ}^d}\left(\frac{L-i}{\eta P_S L/C_1}\right) & \text{if } i < \tau \\ (1 - q_c)\left(1 - F_{H_{SJ}^d}\left(\frac{L-i}{\eta P_S L/C_1}\right)\right) & \text{if } i \geq \tau \end{cases}
\end{aligned} \tag{2.2.102}
$$

6. *The battery is discharged* ($S_j \rightarrow S_i : 0 \leq i < j \leq L$):

$$
\begin{aligned}
p'_{j,i} &= \Pr\{\varepsilon_j \geq \varepsilon_t\}\Pr\{C_{SD} \geq R_s\}\Pr\{\varepsilon_t = \varepsilon_j - \varepsilon_i\} \\
&= \begin{cases} q_c & \text{if } i = j - \tau \\ 0 & \text{otherwise} \end{cases}
\end{aligned} \tag{2.2.103}
$$

Based on the above expressions for $p'_{i,j}$, we define the transition matrix of the above MC as $\mathbf{M}_{HD} := (p'_{i,j})$. Similar as (2.2.75), the stationary distribution of the battery at J' is given by

$$\xi_{HD} = (\mathbf{M}_{HD}^T - \mathbf{I} + \mathbf{B})^{-1}\mathbf{b}. \tag{2.2.104}$$

where the i-th entry, $\xi_{HD,i}$, denotes the probability of the residual energy in the battery of J' being ε_i. As a result, the probability for the energy condition being met at J' is

$$\Pr\{\varepsilon[k] \geq E_{th}\} = \sum_{i=\tau}^{L} \xi_{HD,i}. \tag{2.2.105}$$

2.2.5.2 Secrecy Performance for HD Jammer

In this subsection, we derive the secrecy performance for the cooperative jamming scheme with J'. Based on Proposition 2.3 and Corollary 2.2, we can obtain the following corollaries.

Corollary 2.4 *The closed-form expression of the secrecy outage probability for cooperative jamming from J' can be obtained by replacing N_t and $\xi_{FD,i}$ in (2.2.86) with N_J and $\xi_{HD,i}$, respectively.*

Corollary 2.5 *The closed-form expression of the probability of non-zero secrecy capacity for cooperative jamming from J' can be obtained by replacing N_t in (2.2.89) with N_J, and replacing $\xi_{FD,i}$ with $\xi_{HD,i}$.*

2.2.6 Numerical Results

In this section, We provide numerical results based on the analytical expressions developed in the previous sections, and investigate the impact of key system parameters on the performance. In line with [9], the simulation is carried out on of a linear topology where nodes S, J, E, and D are placed in order along a horizontal line; the distances are set to $d_{SJ} = 5$ m, $d_{SE} = 20$ m and $d_{SD} = 30$ m. Throughout this section, unless otherwise stated, we set the path loss exponent $\alpha = 3$, the fading channel variances $\Omega_{ij} = 1/(1 + d_{ij}^\alpha)$, the noise power $\sigma_D^2 = -80$ dBm, the target secrecy rate $R_s = 1$, the Rician factor $K = 5$ dB, the channel estimation factor $\rho = 1$, and the number of antennas at the jammer $N_J = 8$ (i.e., $N_r = N_t = 4$). For parameters regarding the energy storage, we set the energy conversion efficiency $\eta = 0.5$, the energy transfer efficiency $\eta' = 0.9$, the PES capacity $C_1 = 0.02$, the

SES capacity $C_2 = 0.01$, the discretization level $L = 100$, and the constant circuitry power $P_c = 0.1 \times 10^{-3}$ watt.[10]

2.2.6.1 The Validation of Energy Discretization Model

In this subsection, we examine the accuracy of the energy discretization model (referred to as EDM hereafter, for notation simplicity) presented in Sect. 2.2.2.2. Figure 2.9 shows the secrecy outage probability obtained from (2.2.86) with different PES capacity values and discretization levels. The performance of the continuous energy storage with infinite capacity obtained from (2.2.95) is also plotted to serve as an upper bound. In the case of $C_1 = 0.1$, it can be seen from the figure that the performance of EDM approaches the upper bound as L increases. Specifically, when $L = 400$, the performance of EDM coincides with the upper bound. This is because a larger L results in a smaller quantization step size, i.e., C_1/L, for a given PES capacity C_1. As a result, the energy loss caused by the discretization process reduces. On the other hand, when $C_1 = 0.02$, it is observed that the performance of EDM converges to the upper bound much more rapidly than the case of $C_1 = 0.1$. In particular, even a small discretization level of $L = 50$ suffices the close match. This is because given a small C_1, a small value of L is adequate to provide the same discretization granularity. This observation allows the system designer to reduce computation via choosing a small L, when the energy

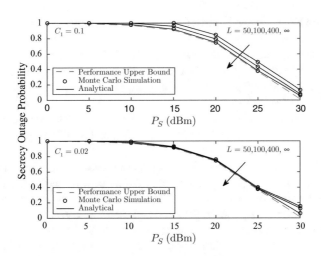

Fig. 2.9 Secrecy outage probability with various C_1 and L versus P_S. $P_J = 10$ dBm

---

[10]We note that typical values for practical parameters used in EH systems depend on both the system application and specific technology used for implementation of RF energy harvesting circuits.

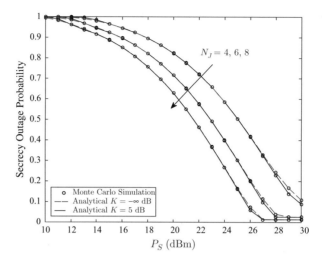

Fig. 2.10 Secrecy outage probability with various N_J and K versus P_S. $P_J = 0$ dBm

storage capacity is low. Besides, when P_S exceeds 25 dBm, the performance of EDM deviates from the upper bound, which indicates that energy overflow occurs frequently, and the selected storage capacity should be enlarged.

2.2.6.2 The Effect of the Number of Jammer Antennas and Rician Factor

In this subsection, we investigate the effects of the number of antennas at the jammer (i.e., N_J) and the Rician factor (i.e., K) on the secrecy outage probability derived in (2.2.86). In Fig. 2.10, the solid lines are for $K = 5$ dB (i.e., Rician fading), whereas the dashed lines are for $K = -\infty$ dB (i.e., Rayleigh fading). The performance differences between these two are surprisingly minor, which indicates that the strength of the LoS path between S and J has limited impact on the system performance. On the contrary, the effect of N_J is remarkable: As N_J increases from 4 to 8, the secrecy outage decreases significantly. In addition, the increase of P_S also improves the performance notably. The positive association of N_J and P_S with system performance is because greater N_J and/or P_S can increase the amount of harvested energy at the jammer, and therefore can support more frequent jamming. The finding suggests that increasing the number of antennas at the jammer and/or increasing the transmitting power at the source are two effective manners for secrecy improvement. Monte Carlo simulation results are also provided in Fig. 2.10 to validate the closed-form expressions in Eq. (2.2.86). In addition, similar positive effects of N_J and K on the probability of non-zero secrecy capacity derived in (2.2.89) can be observed in Fig. 2.11. Monte Carlo simulation results presented in Fig. 2.11 are in line with the closed-form expressions in Eq. (2.2.89).

Fig. 2.11 The existence of non-zero secrecy capacity with various N_J and K versus P_S. $P_J = 0$ dBm

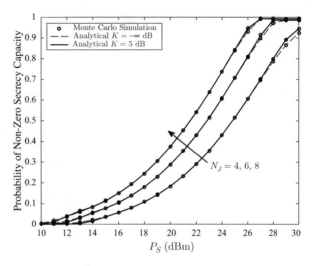

Fig. 2.12 Secrecy outage probability with various P_S versus jamming power P_J

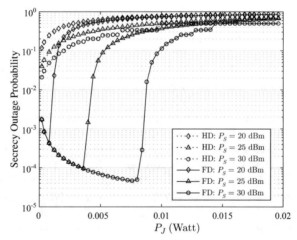

2.2.6.3   The Effect of the Jamming Power

Figure 2.12 shows the association between the secrecy outage probability and the jamming power P_J. The source transmitting power is chosen from $P_S = [20, 25, 30]$ dBm. Overall, it can be seen that a distinct optimum jamming power P_J^* with the minimum secrecy outage probability, exists in all considered scenarios. The existence of P_J^* is because, in short, a higher P_J is associated with lower jamming frequency but higher interference strength. This finding validates our deduction in Remark 2.4. It also implies that in a scenario with multiple jammers, the jamming power at each jammer should be individually optimized. In addition, as expected, the optimum jamming power of the proposed FD scheme is notably higher than that of HD.

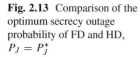

Fig. 2.13 Comparison of the optimum secrecy outage probability of FD and HD, $P_J = P_J^*$

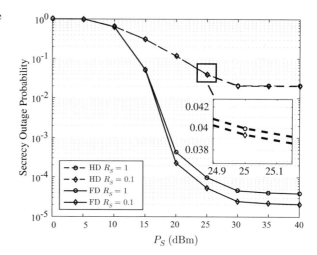

2.2.6.4   The Performance Comparison Between FD Jamming and HD Jamming

In this subsection, we compare the optimum secrecy performance between the FD and the HD scheme. Figure 2.13 illustrates the secrecy outage probability for the two schemes with various secrecy rate R_s. The jamming power for both FD and HD is chosen to be the corresponding optimum value, i.e., $P_J = P_J^*$. It is clear from the figure that the proposed FD scheme achieves significantly lower secrecy outage than the HD scheme over the entire range of P_S. Specifically, when R_s reduces from 1 to 0.1, the reduction in secrecy outage for the FD scheme is more notable than that for the HD scheme, suggesting that reducing R_s as a method to mitigate the outage is more effective in the FD scheme than in the HD scheme. Moreover, the performance gap between the two schemes can be further enlarged by rearranging the antenna allocation at the FD jammer, which will be discussed in the next subsection.

2.2.6.5   The Effect of Antenna Allocation at the Jammer

In this subsection, we investigate the impact of antenna allocation at the jammer on system performance. Figure 2.14 shows the secrecy outage probability of the proposed protocol with different transmitting/receiving antenna allocations. When P_S increases from 10 to 15 dBm, the allocation of $N_t = 2$, $N_r = 6$ achieves the smallest secrecy outage; after 15 dBm, equal allocation of $N_t = 4$, $N_r = 4$ overtakes until P_S increases to 35 dBm. In the high transmitting power regime, the allocation of $N_t = 6$, $N_r = 2$ finally catches up. The finding suggests that in the cases that the source is sending with low transmitting power, more antennas should be used for energy harvesting, whereas in the paradigm where the source is sending with high transmitting power, more antennas should be used for cooperative jamming, as fewer antennas are required to receive sufficient energy.

Fig. 2.14 Secrecy outage
probability with various
antenna allocation at the
jammer, $P_J = P_J^*$

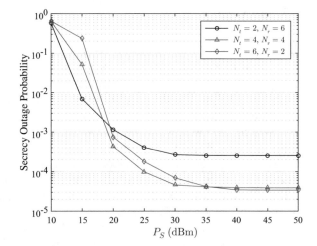

Fig. 2.15 Secrecy outage
probability with various ρ
versus P_S, $P_J = P_J^*$

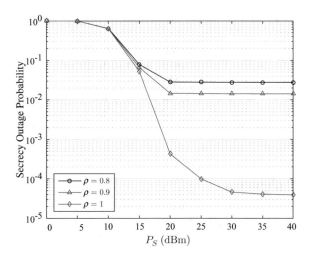

2.2.6.6   The Effect of Channel Estimation Error

In this subsection, we investigate the impact of ρ on the secrecy outage probability of the proposed protocol. From Fig. 2.15, as expected, the CSI mismatch
indeed results in performance loss. Specifically, the performance loss is dramatic
when slightly reducing ρ from 1 to 0.9. The finding indicates that the system
performance in practice can be severely degraded by imperfect CSI. Therefore,
developing advanced CSI estimation technique dedicatedly for wireless-powered
communication network is critical for physical layer security.

2.3 Summary

This chapter started with studying the secure transmission from a wireless-powered source to a destination with the help of a multi-antenna PB, in the presence of multiple eavesdroppers. A secure transmission scheme named ATT was proposed where PB acted as a power supply during the WPT mode and as a cooperative jammer during the IT mode. The resultant energy state transitions of the source were investigated through an energy discretization method and a finite-state Markov Chain. Additionally, the closed-form expressions were derived for three performance metrics, including the connection outage probability, the secrecy outage probability, and the secrecy throughput. Numerical results validated our theoretical analysis, investigated the impacts of various system parameters, and revealed the merits of our proposed ATT scheme over a benchmark HTT protocol.

In addition, the use of a wireless-powered FD jammer has also been exploited to provide PLS for wireless channels with imperfect CSI. The impact of the channel estimation error was analyzed on the secrecy performance of the proposed AnJ protocol based on secrecy outage probability. The results revealed the importance of the carefully chosen jamming power in this scenario. Regarding the state transitions taking place at the jammer, we have obtained its long-term stationary energy state distribution. For comparison, an ideal energy storage model with infinite capacity and a wireless-powered HD jammer were also investigated. Finally, numerical results demonstrated that our proposed protocol can not only provide a superior performance over the conventional HD jammer, but also a satisfactory performance close to the upper bound when the energy storage is sufficiently subdivided.

References

1. L. Tang and Q. Li, "Wireless Power Transfer and Cooperative Jamming for Secrecy Throughput Maximization," *IEEE Wireless Commun. Lett.*, vol. PP, no. 99, pp. 1–1, 2016.
2. X. Jiang, C. Zhong, Z. Zhang, and G. K. Karagiannidis, "Power Beacon Assisted Wiretap Channels With Jamming," *IEEE Trans. Wireless Commun.*, vol. 15, no. 12, pp. 8353–8367, Dec. 2016.
3. A. Wyner, "The wire-tap channel," *Bell Sys. Tech. J.*, vol. 54, no. 8, pp. 1355–1387, Oct. 1975.
4. J. Barros and M. Rodrigues, "Secrecy Capacity of Wireless Channels," in *Proc. IEEE Int. Symp. Inf. Theory*, Seattle, WA, USA, 2006, pp. 356–360.
5. R. Negi and S. Goel, "Secret communication using artificial noise," in *in Proc. IEEE VTC*, Dallas, TX, USA, 2005, pp. 1906–1910.
6. X. Zhou and M. McKay, "Secure Transmission With Artificial Noise Over Fading Channels: Achievable Rate and Optimal Power Allocation," *IEEE Trans. Veh. Technol.*, vol. 59, no. 8, pp. 3831–3842, Oct. 2010.
7. S. Goel and R. Negi, "Guaranteeing Secrecy using Artificial Noise," *IEEE Trans. Wireless Commun.*, vol. 7, no. 6, pp. 2180–2189, Jun. 2008.
8. W. Mou, Y. Cai, W. Yang, W. Yang, X. Xu, and J. Hu, "Exploiting full Duplex techniques for secure communication in SWIPT system," in *Proc. WCSP*, Nanjing, China, Oct. 2015, pp. 1–6.

9. W. Liu, X. Zhou, S. Durrani, and P. Popovski, "Secure Communication with a Wireless-Powered Friendly Jammer," *IEEE Trans. Wireless Commun.*, vol. 15, no. 1, pp. 401–415, Jan. 2016.

10. X. Jiang, C. Zhong, X. Chen, T. Q. Duong, T. A. Tsiftsis, and Z. Zhang, "Secrecy Performance of Wirelessly Powered Wiretap Channels," *IEEE Trans. Commun.*, vol. 64, no. 9, pp. 3858–3871, Sep. 2016.

11. H. Xing, K.-K. Wong, Z. Chu, and A. Nallanathan, "To Harvest and Jam: A Paradigm of Self-Sustaining Friendly Jammers for Secure AF Relaying," *IEEE Trans. Signal Process.*, vol. 63, no. 24, pp. 6616–6631, Dec. 2015.

12. I. Krikidis, S. Timotheou, and S. Sasaki, "RF Energy Transfer for Cooperative Networks: Data Relaying or Energy Harvesting?" *IEEE Commun. Lett.*, vol. 16, no. 11, pp. 1772–1775, Nov. 2012.

13. M. Haenggi, *Stochastic Geometry for Wireless Networks* .Cambridge University Press, Oct. 2012.

14. Z. Ding and H. V. Poor, "Cooperative Energy Harvesting Networks With Spatially Random Users," *IEEE Signal Process. Lett.*, vol. 20, no. 12, pp. 1211–1214, Dec. 2013.

15. P. Liu, S. Gazor, I. M. Kim, and D. I. Kim, "Noncoherent Relaying in Energy Harvesting Communication Systems," *IEEE Trans. Wireless Commun.*, vol. 14, no. 12, pp. 6940–6954, Dec. 2015.

16. I. Krikidis, "SWIPT in 3-D Bipolar Ad Hoc Networks With Sectorized Antennas," *IEEE Commun. Lett.*, vol. 20, no. 6, pp. 1267–1270, Jun. 2016.

17. S. Luo, J. Li, and A. Petropulu, "Uncoordinated cooperative jamming for secret communications," *IEEE Trans. Inf. Forensics Security*, vol. 8, no. 7, pp. 1081–1090, July 2013.

18. A. Thangaraj, S. Dihidar, A. R. Calderbank, S. W. McLaughlin, and J. M. Merolla, "Applications of LDPC Codes to the Wiretap Channel," *IEEE Trans. Inf. Theory*, vol. 53, no. 8, pp. 2933–2945, Aug. 2007.

19. J. Xu and R. Zhang, "Energy Beamforming With One-Bit Feedback," *IEEE Trans. Signal Process.*, vol. 62, no. 20, pp. 5370–5381, Oct. 2014.

20. H. J. Visser and R. J. M. Vullers, "RF Energy Harvesting and Transport for Wireless Sensor Network Applications: Principles and Requirements," vol. 101, no. 6, pp. 1410–1423, Jun. 2013.

21. A. Khaligh and Z. Li, "Battery, Ultracapacitor, Fuel Cell, and Hybrid Energy Storage Systems for Electric, Hybrid Electric, Fuel Cell, and Plug-In Hybrid Electric Vehicles: State of the Art," *IEEE Trans. Veh. Technol.*, vol. 59, no. 6, pp. 2806–2814, Jul. 2010.

22. X. Lu, P. Wang, D. Niyato, D. I. Kim, and Z. Han, "Wireless Networks With RF Energy Harvesting: A Contemporary Survey," *IEEE Commun. Surveys Tuts.*, vol. 17, no. 2, pp. 757–789, 2015.

23. M. Maso, C. F. Liu, C. H. Lee, T. Q. S. Quek, and L. S. Cardoso, "Energy-Recycling Full-Duplex Radios for Next-Generation Networks," *IEEE J. Sel. Areas Commun.*, vol. 33, no. 12, pp. 2948–2962, Dec. 2015.

24. J. Agust, G. Abadal, and J. Alda, "Electromagnetic radiation energy investing — the rectenna based approach," in *ICT — Energy-Concepts Towards Zero-Power Information and Communication Technology*, InTech, Feb. 2014. [Online]. Available: http://www.intechopen.com/books/ict-energy-concepts-towards-zero-power-information-and-communication-technology/electromagnetic-radiation-energy-harvesting-the-rectenna-based-approach

25. Y. Zeng and R. Zhang, "Full-Duplex Wireless-Powered Relay With Self-Energy Recycling," *IEEE Wireless Commun. Lett.*, vol. 4, no. 2, pp. 201–204, Apr. 2015.

26. C. Zhong, H. Suraweera, G. Zheng, I. Krikidis, and Z. Zhang, "Wireless Information and Power Transfer With Full Duplex Relaying," *IEEE Trans. Commun.*, vol. 62, no. 10, pp. 3447–3461, Oct. 2014.

27. Y. Che, J. Xu, L. Duan, and R. Zhang, "Multiantenna Wireless Powered Communication With Cochannel Energy and Information Transfer," *IEEE Commun. Lett.*, vol. 19, no. 12, pp. 2266–2269, Dec. 2015.

28. J. Xu, S. Bi, and R. Zhang, "Multiuser MIMO Wireless Energy Transfer With Coexisting Opportunistic Communication," *IEEE Wireless Commun. Lett.*, vol. 4, no. 3, pp. 273–276, Jun. 2015.
29. Y. Zeng and R. Zhang, "Optimized Training Design for Wireless Energy Transfer," *IEEE Trans. Commun.*, vol. 63, no. 2, pp. 536–550, Feb. 2015.
30. W. Feng, Y. Wang, N. Ge, J. Lu and J. Zhang, "Virtual MIMO in Multi-Cell Distributed Antenna Systems: Coordinated Transmissions with Large-Scale CSIT," *IEEE J. Sel. Areas Commun.*, vol. 31, no. 10, pp. 2067–2081, Oct 2013.
31. W. Feng, Y. Wang, D. Lin, N. Ge, J. Lu and S. Li, "When mmWave Communications Meet Network Densification: A Scalable Interference Coordination Perspective," *IEEE J. Sel. Areas Commun.*, vol. 35, no. 7, pp. 1459–1471, Jul 2017.
32. X. Chen, J. Chen, H. Zhang, Y. Zhang, and C. Yuen, "On Secrecy Performance of A Multi-Antenna Jammer Aided Secure Communications with Imperfect CSI," *IEEE Trans. Veh. Technol.*, vol. PP, no. 99, pp. 1–1, 2015.
33. D. S. Michalopoulos, H. A. Suraweera, G. K. Karagiannidis, and R. Schober, "Amplify-and-Forward Relay Selection with Outdated Channel Estimates," *IEEE Trans. Commun.*, vol. 60, no. 5, pp. 1278–1290, May 2012.
34. Q. Shi, C. Peng, W. Xu, M. Hong, and Y. Cai, "Energy Efficiency Optimization for MISO SWIPT Systems With Zero-Forcing Beamforming," *IEEE Trans. Signal Process.*, vol. 64, no. 4, pp. 842–854, Feb. 2016.
35. S. Cui, A. J. Goldsmith, and A. Bahai, "Energy-efficiency of MIMO and cooperative MIMO techniques in sensor networks," *IEEE J. Sel. Areas Commun.*, vol. 22, no. 6, pp. 1089–1098, Aug. 2004.
36. W.-J. Huang, Y.-W. Hong, and C.-C. Kuo, "Lifetime maximization for amplify-and-forward cooperative networks," *IEEE Trans. Wireless Commun.*, vol. 7, no. 5, pp. 1800–1805, May 2008.
37. Y.-c. Ko, A. Abdi, M.-S. Alouini, and M. Kaveh, "Average outage duration of diversity systems over generalized fading channels," in *Proc. IEEE WCNC*, Chicago, Il, USA, 2000, pp. 216–221.
38. I. S. Gradshteyn and I. M. Ryzhik, *Table of integrals, series, and products*, 7th ed. New York: Academic Press, 2007.
39. I. Krikidis, T. Charalambous, and J. Thompson, "Buffer-Aided Relay Selection for Cooperative Diversity Systems without Delay Constraints," *IEEE Trans. Wireless Commun.*, vol. 11, no. 5, pp. 1957–1967, May 2012.

Chapter 3
Extending Wireless Powered Communication Networks for Future Internet of Things

Abstract In this chapter, we propose a DH-WPCN with one HAP and a number of relays and users. Assuming that the users have fixed energy supplies and the relays need to harvest energy from RF transmission of the HAP, we presented uplink and downlink communication protocols. Optimal values of parameters for maximizing the total throughput of the network in both directions were found. Specifically, we formulated uplink and downlink sum-throughput maximization problems to find optimal time allocation in both uplink and downlink communications as well as optimal power splitting factors in downlink communication. The convex structure of the uplink throughput maximization problem allowed us to obtain the optimal value of the time-slot durations for energy and information transfer in closed form, while in downlink throughput maximization, we used iterations for finding a near-optimal solution due to the non-convexity of the problem. We evaluated the uplink and downlink throughput performance of our proposed schemes via simulations and identified the existence of the doubly near-far problem in uplink communication which results in extremely unfair throughput distribution among the users. Due to the dependence of each user's achievable throughput on its corresponding relay's distance from the HAP, we proposed to dynamically adjust the location of the relays to attain a more balanced throughput allocation in the network. Numerical results confirmed that the position of the relays has a major impact on users' throughputs and the severity of the throughput unfairness can be controlled by changing relays' placements. Next in Sect. 3.2, we investigate a more robust solution for tackling the aforementioned fairness issue and developed a fairness enhancement scheme to provide all users with equal throughput. We formulated a minimum throughput maximization problem and proposed a novel algorithm for finding the maximum common throughput of the users plus the optimal time allocation for achieving the maximum level of fairness. We also conducted simulations to compare the performance of the proposed fairness-improving scheme with the strategy presented in Sect. 3.1. Simulation results revealed a throughput-fairness trade-off in our DH-WPCN implying that the fairness is achieved at the cost of total throughput reduction. Therefore, depending on the network requirements in terms of sum-throughput and fairness, either the strategy proposed in Sect. 3.1 or the scheme presented in Sect. 3.2 can be chosen as the optimal policy.

© Springer Nature Switzerland AG 2019

A. Jamalipour, Y. Bi, *Wireless Powered Communication Networks*,

https://doi.org/10.1007/978-3-319-98174-1_3

3.1   Throughput Maximization in DH-WPCN

3.1.1   System Model

As shown in Fig. 3.1, we consider a DH-WPCN with one HAP, K users, and K energy-constrained relays. HAP and users are assumed to have stable energy sources. There is no direct link between the HAP and the users and the ith relay is responsible for forwarding the ith user's data to the HAP and vice versa,[1] but it needs to harvest energy before assisting the communication. HAP, users, and relays are all equipped with one single antenna each.

Without loss of generality, we assume that channel reciprocity holds for all channels. The channel coefficients between the HAP and the relay R_i and between R_i and the user U_i are denoted by h_i and g_i, respectively. All channels are quasi-static flat fading and remain constant over a transmission block, but can change independently from one block to another. It is further assumed that h_i and g_i are perfectly known at the HAP and U_i.

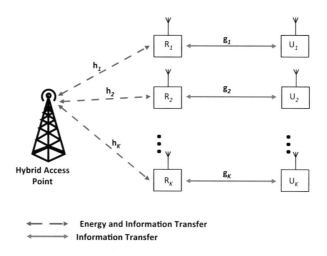

Fig. 3.1 A dual-hop wireless powered communication network

---

[1]The relays can also cooperatively serve the scheduled users, however, we assume a single relay responsible for each user because it reduces the complexity of coordination and implementation [68]. Furthermore, the number of relays can be different from the number of users, but if there are less relays in the network than users, some relays will have to serve more than one user which reduces the energy allocated for information transmission from/to each user and consequently the throughput will be degraded. It is also obvious that if the number of relays is greater than the number of users, some of the relays will remain idle during a whole transmission block which is not efficient. For these reasons, we assume a dedicated relay for each user.

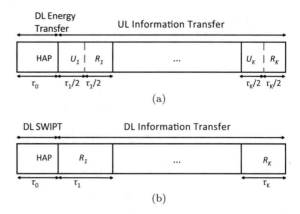

Fig. 3.2 Frame structures for dual-hop wireless powered communication network. (**a**) Uplink communication frame structure. (**b**) Downlink communication frame structure

The frame structures for uplink and downlink communications are illustrated in Fig. 3.2a and b, respectively. For convenience, normalized transmission blocks are used throughout this chapter. Also, superscripts u and d are used to differentiate between uplink and downlink communications.

3.1.1.1 Uplink Communication Model

In the first τ_0 amount of time ($0 < \tau_0 < 1$), the HAP broadcasts an energizing signal with a constant transmit power P_A to all relays to power them for their upcoming forwarding tasks. Uplink information transmission from U_i to the HAP takes place in the ith time slot ($i = 1, \ldots, K$) with $\tau_i/2$ amount of time given to U_i to transmit its information to R_i and the remaining $\tau_i/2$ devoted to R_i to forward U_i's data to the HAP. The energy harvested by R_i during τ_0 can be obtained as

$$E_{r,i}^u = \eta_i P_A |h_i|^2 \tau_0, \qquad (3.1.1)$$

where $0 < \eta_i < 1$ ($i = 1 \ldots, K$) is the energy harvesting efficiency at R_i. Assuming all this energy is used up for forwarding U_i's data during the second half of the ith time slot, the average transmit power of R_i is given by

$$P_{r,i}^u = \frac{E_{r,i}^u}{\dfrac{\tau_i}{2}} = 2\eta_i P_A |h_i|^2 \frac{\tau_0}{\tau_i}. \qquad (3.1.2)$$

We denote $x_i^u \sim \mathcal{CN}(0,1)$ as the signal transmitted by U_i during the first half of the ith time slot, where $\mathcal{CN}(\mu, \sigma^2)$ stands for a circularly symmetric complex Gaussian random variable with mean μ and variance σ^2. The sampled baseband signal at R_i can be written as

$$y_{r,i}^u = g_i \sqrt{P_{U,i}} x_i^u + z_{r,i}, \qquad (3.1.3)$$

where $P_{U,i}$ represents the transmit power of U_i and $z_{r,i} \sim \mathcal{CN}(0, \sigma_{r,i}^2)$ indicates the overall additive Gaussian noise at R_i due to the receiving antenna and RF band to baseband signal conversion. R_i amplifies the received signal and forwards it to the HAP. The signal received at the HAP in the ith slot is thus given by

$$
y_{H,i}^u = \frac{h_i \sqrt{P_{r,i}^u} (g_i \sqrt{P_{U,i}} x_i^u + z_{r,i})}{\sqrt{|g_i|^2 P_{U,i} + \sigma_{r,i}^2}} + z_H, \tag{3.1.4}
$$

where the factor $\sqrt{|g_i|^2 P_{U,i} + \sigma_{r,i}^2}$ in the denominator is the power constraint factor at R_i [27] and $z_H \sim \mathcal{CN}(0, \sigma_H^2)$ is the additive Gaussian noise at the HAP. Replacing (3.1.2) into (3.1.4), we will have

$$
\begin{aligned}
y_{H,i}^u = &\left(\frac{h_i g_i \sqrt{P_{U,i}} \sqrt{2\eta_i P_A |h_i|^2 \frac{\tau_0}{\tau_i}}}{\sqrt{|g_i|^2 P_{U,i} + \sigma_{r,i}^2}} x_i^u \right) \\
&+ \left(\frac{h_i \sqrt{2\eta_i P_A |h_i|^2 \frac{\tau_0}{\tau_i}}}{\sqrt{|g_i|^2 P_{U,i} + \sigma_{r,i}^2}} z_{r,i} + z_H \right),
\end{aligned} \tag{3.1.5}
$$

where the expression within the first parentheses indicates the signal part and the second parentheses include the noise part. Therefore, signal-to-noise ratio (SNR) can be expressed as

$$
\gamma_i^u = \frac{\dfrac{2\eta_i P_A P_{U,i} |h_i|^4 |g_i|^2 \frac{\tau_0}{\tau_i}}{|g_i|^2 P_{U,i} + \sigma_{r,i}^2}}{\dfrac{2\eta_i P_A |h_i|^4 \sigma_{r,i}^2 \frac{\tau_0}{\tau_i}}{|g_i|^2 P_{U,i} + \sigma_{r,i}^2} + \sigma_H^2} \simeq \frac{2\eta_i P_A P_{U,i} |h_i|^4 |g_i|^2 \frac{\tau_0}{\tau_i}}{2\eta_i P_A |h_i|^4 \sigma_{r,i}^2 \frac{\tau_0}{\tau_i} + P_{U,i} |g_i|^2 \sigma_H^2}. \tag{3.1.6}
$$

Note that we have neglected the term $\sigma_H^2 . \sigma_{r,i}^2$ in the denominator as its amount is insignificant compared to $P_{U,i} |g_i|^2 \sigma_H^2$. Finally, the achievable uplink throughput from U_i to the HAP is given by

$$
\mathcal{T}_i^u = \frac{\tau_i}{2} \log(1 + \gamma_i^u) \simeq \frac{\tau_i}{2} \log\left(1 + \frac{A_i^u \frac{\tau_0}{\tau_i}}{B_i^u \frac{\tau_0}{\tau_i} + C_i^u} \right), \tag{3.1.7}
$$

where

$$A_i^u = 2\eta_i P_A P_{U,i} |h_i|^4 |g_i|^2, \tag{3.1.8}$$

$$B_i^u = 2\eta_i P_A |h_i|^4 \sigma_{r,i}^2, \tag{3.1.9}$$

$$C_i^u = P_{U,i} |g_i|^2 \sigma_H^2. \tag{3.1.10}$$

3.1.1.2 Downlink Communication Model

In the 0th slot (with duration τ_0), the HAP broadcasts a signal to all relays which carries both information and energy. Exploiting a power splitting method, R_i uses the proportion ρ_i of the received signal power for harvesting energy which will be used in the ith time slot (with duration τ_i) for forwarding HAP's information to U_i. Therefore, the harvested energy and the average transmit power of R_i are expressed as

$$E_{r,i}^d = \eta_i \rho_i P_A |h_i|^2 \tau_0, \tag{3.1.11}$$

$$P_{r,i}^d = \eta_i \rho_i P_A |h_i|^2 \frac{\tau_0}{\tau_i}, \tag{3.1.12}$$

respectively. The signal with the remaining $(1 - \rho_i)$ proportion of the initial power will be amplified and forwarded to U_i. Denoting $x_H \sim \mathcal{CN}(0,1)$ as the signal transmitted by the HAP during τ_0, the sampled baseband signal at R_i is given by

$$y_{r,i}^d = h_i \sqrt{(1 - \rho_i) P_A} x_H + z_{r,i}. \tag{3.1.13}$$

After the AF process, the received signal at U_i is expressed as[2]

$$y_{U,i}^d = \frac{g_i \sqrt{P_{r,i}^d} (h_i \sqrt{(1 - \rho_i) P_A} x_H + z_{r,i})}{\sqrt{|h_i|^2 (1 - \rho_i) P_A + \sigma_{r,i}^2}} + z_{U,i}, \tag{3.1.14}$$

where $z_{U,i} \sim \mathcal{CN}(0, \sigma_{U,i}^2)$ is the additive Gaussian noise at U_i. Following the same steps as in uplink communication, SNR at U_i is obtained as

$$\gamma_i^d \simeq \frac{\eta_i P_A^2 |h_i|^4 |g_i|^2 \rho_i (1 - \rho_i) \frac{\tau_0}{\tau_i}}{\eta_i P_A |h_i|^2 |g_i|^2 \sigma_{r,i}^2 \rho_i \frac{\tau_0}{\tau_i} + P_A |h_i|^2 \sigma_{U,i}^2 (1 - \rho_i)}, \tag{3.1.15}$$

---

[2]It is assumed that the information forwarded by the ith relay to the ith user cannot be properly decoded at the jth user ($j \neq i$); so, each user only decodes the information received from its dedicated relay.

and the achievable downlink throughput from the HAP to U_i is given by

$$T_i^d = \tau_i \log(1 + \gamma_i^d) \simeq \tau_i \log(1 + \frac{A_i^d \rho_i (1 - \rho_i) \frac{\tau_0}{\tau_i}}{B_i^d \rho_i \frac{\tau_0}{\tau_i} + C_i^d (1 - \rho_i)}), \qquad (3.1.16)$$

where

$$A_i^d = \eta_i P_A^2 |h_i|^4 |g_i|^2, \qquad (3.1.17)$$

$$B_i^d = \eta_i P_A |h_i|^2 |g_i|^2 \sigma_{r,i}^2, \qquad (3.1.18)$$

$$C_i^d = P_A |h_i|^2 \sigma_{U,i}^2. \qquad (3.1.19)$$

3.1.2　Uplink Throughput Maximization

In this section, we are interested in finding optimal time allocations for energy transfer and information transmission in uplink communication in order to maximize the total uplink throughput. According to (3.1.7), uplink throughput maximization problem is formulated as

$$(P1) : \max_{\tau_0, \tau_i} \sum_{i=1}^{K} T_i^u (\tau_0, \tau_i) =$$

$$\max_{\tau_0, \tau_i} \frac{1}{2} \sum_{i=1}^{K} \tau_i \log(1 + \frac{A_i^u \frac{\tau_0}{\tau_i}}{B_i^u \frac{\tau_0}{\tau_i} + C_i^u}), \qquad (3.1.20)$$

$$s.t. \sum_{i=0}^{K} \tau_i \leqslant 1, \qquad (3.1.21)$$

$$0 \leqslant \tau_i \leqslant 1, \quad i = 0, \ldots, K. \qquad (3.1.22)$$

The constraint in (3.1.21) is the time constraint implying that sum of time durations for energy and information transfer must not exceed the length of the transmission block which is assumed to be 1.

Lemma 3.1 $T_i^u (\tau_0, \tau_i)$ *is concave for* $0 \leqslant \frac{\tau_0}{\tau_i} \leqslant \infty$.

Proof $T_i^u (\tau_0, \tau_i) = \frac{\tau_i}{2} \log(1 + \frac{A_i^u \tau_0/\tau_i}{B_i^u \tau_0/\tau_i + C_i^u})$ *is a perspective function of* $f(\tau_0) = \frac{1}{2} \log(1 + \frac{A_i^u \tau_0}{B_i^u \tau_0 + C_i^u})$ *and* $f(\tau_0)$ *is concave over* $\tau_0 \geqslant 0$. *Since the perspective* operation preserves concavity, $T_i^u(\tau_0, \tau_i)$ is a jointly concave function of (τ_0, τ_i). This completes the proof of Lemma 3.1. ∎

A non-negative weighted summation of concave functions is also concave. Hence, the objective function of $P1$ is a concave function of $\tau = [\tau_0, \tau_1, \ldots, \tau_K]$. Moreover, the constraints of $P1$ are affine. Therefore, $P1$ is convex and can be solved using convex optimization techniques [69]. Applying Lagrange duality method, we have

$$\mathcal{L}(\tau, \lambda) = \frac{1}{2} \sum_{i=1}^{K} \tau_i \log(1 + \frac{A_i^u \frac{\tau_0}{\tau_i}}{B_i^u \frac{\tau_0}{\tau_i} + C_i^u}) - \lambda(\sum_{i=0}^{K} \tau_i - 1), \quad (3.1.23)$$

where $\lambda \geqslant 0$ is the Lagrangian multiplier associated with the constraint in (3.1.21). Then, the dual function is written as

$$\mathcal{G}(\lambda) = \max_{D} \mathcal{L}(\tau, \lambda), \quad (3.1.24)$$

where D is the feasible set of τ determined by (3.1.21) and (3.1.22). The optimal solution of $P1$ can be obtained from the following theorem.

Theorem 3.1 *Optimal time allocations of $P1$ are given by*

$$\tau_0^* = \frac{1}{1 + \sum_{j=1}^{K} \frac{1}{\zeta_j^*}}, \quad (3.1.25)$$

$$\tau_i^* = \frac{1}{\zeta_i^*(1 + \sum_{j=1}^{K} \frac{1}{\zeta_j^*})}, \quad (3.1.26)$$

where $\zeta_i^*, i = 1, \ldots, K$ is the solution of

$$\frac{1}{2}\left[\log(1 + \frac{A_i^u \zeta_i}{B_i^u \zeta_i + C_i^u}) - \frac{A_i^u C_i^u \zeta_i}{(B_i^u \zeta_i + C_i^u)(A_i^u \zeta_i + B_i^u \zeta_i + C_i^u)}\right] = \lambda^*, \quad (3.1.27)$$

and λ^* is the optimal dual solution.

Proof As we can find a $\tau \in D$ such that $\tau_i > 0$ $(i = 0, 1, \ldots, K)$ and $\sum_{i=0}^{K} \tau_i < 1$, $P1$ satisfies Slater's condition and the duality gap is zero [69]. As a result, $P1$ can be solved by applying Karush-Kuhn-Tucker (KKT) conditions:

$$\frac{\partial \mathcal{L}(\tau^*, \lambda^*)}{\partial \tau_0} = 0, \quad (3.1.28)$$

$$\frac{\partial \mathcal{L}(\tau^*, \lambda^*)}{\partial \tau_i} = 0, \tag{3.1.29}$$

$$\sum_{i=0}^{K} \tau_i^* \leqslant 1, \tag{3.1.30}$$

$$\lambda^*(\sum_{i=0}^{K} \tau_i^* - 1) = 0. \tag{3.1.31}$$

From (3.1.28) and (3.1.29) it follows that

$$\frac{1}{2} \sum_{i=1}^{K} \frac{A_i^u C_i^u}{(B_i^u \frac{\tau_0^*}{\tau_i^*} + C_i^u)(A_i^u \frac{\tau_0^*}{\tau_i^*} + B_i^u \frac{\tau_0^*}{\tau_i^*} + C_i^u)} = \lambda^*, \tag{3.1.32}$$

$$\frac{1}{2}\left[\log(1 + \frac{A_i^u \frac{\tau_0^*}{\tau_i^*}}{B_i^u \frac{\tau_0^*}{\tau_i^*} + C_i^u}) - \frac{A_i^u C_i^u \frac{\tau_0^*}{\tau_i^*}}{(B_i^u \frac{\tau_0^*}{\tau_i^*} + C_i^u)(A_i^u \frac{\tau_0^*}{\tau_i^*} + B_i^u \frac{\tau_0^*}{\tau_i^*} + C_i^u)} \right] = \lambda^*. \tag{3.1.33}$$

Setting $\zeta_i = \frac{\tau_0^*}{\tau_i^*}$, it is easy to show that the left hand side of (3.1.33) is a monotonically increasing function of ζ_i for $\zeta_i \geqslant 0$ and we have

$$\lim_{\zeta_i \to +\infty} \frac{1}{2}\left[\log(1 + \frac{A_i^u \zeta_i}{B_i^u \zeta_i + C_i^u}) - \frac{A_i^u C_i^u \zeta_i}{(B_i^u \zeta_i + C_i^u)(A_i^u \zeta_i + B_i^u \zeta_i + C_i^u)} \right]$$

$$= \frac{1}{2} \log(1 + \frac{A_i^u}{B_i^u}). \tag{3.1.34}$$

Hence, λ^* is upper-bounded by $\frac{1}{2} \log(1 + \frac{A_i^u}{B_i^u})$ and there exists a unique ζ_i^* satisfying (3.1.33) for $0 \leqslant \lambda^* < \frac{1}{2} \log(1 + \frac{A_i^u}{B_i^u})$. Now, using bisection method, we can find λ^* such that

$$\frac{1}{2}\left[\log(1 + \frac{A_i^u \zeta_i^*}{B_i^u \zeta_i^* + C_i^u}) - \frac{A_i^u C_i^u \zeta_i^*}{(B_i^u \zeta_i^* + C_i^u)(A_i^u \zeta_i^* + B_i^u \zeta_i^* + C_i^u)} \right] = \lambda^*$$

$$\forall i \in \{1, \ldots, K\} \tag{3.1.35}$$

and

$$\frac{1}{2}\sum_{i=1}^{K}\frac{A_i^u C_i^u}{(B_i^u \zeta_i^* + C_i^u)(A_i^u \zeta_i^* + B_i^u \zeta_i^* + C_i^u)} = \lambda^*. \tag{3.1.36}$$

Then, from (3.1.31) we have

$$\tau_0^* + \frac{\tau_0^*}{\zeta_1^*} + \ldots + \frac{\tau_0^*}{\zeta_K^*} = 1. \tag{3.1.37}$$

Therefore

$$\tau_0^* = \frac{1}{1 + \sum_{j=1}^{K}\frac{1}{\zeta_j^*}}, \tag{3.1.38}$$

$$\tau_i^* = \frac{1}{\zeta_i^*(1 + \sum_{j=1}^{K}\frac{1}{\zeta_j^*})}. \tag{3.1.39}$$

This completes the proof of Theorem 3.1. ∎

3.1.3 Downlink Throughput Maximization

In this section, we want to maximize the total downlink throughput by finding the optimal time allocations and power splitting ratios. From (3.1.16), we have the following optimization problem:

$$(P2): \max_{\tau_0, \tau_i, \rho_i} \sum_{i=1}^{K} T_i^d(\tau_0, \tau_i, \rho_i) =$$

$$\max_{\tau_0, \tau_i, \rho_i} \sum_{i=1}^{K} \tau_i \log(1 + \frac{A_i^d \rho_i (1 - \rho_i)\frac{\tau_0}{\tau_i}}{B_i^d \rho_i \frac{\tau_0}{\tau_i} + C_i^d(1 - \rho_i)}), \tag{3.1.40}$$

$$s.t. \sum_{i=0}^{K} \tau_i \leqslant 1, \tag{3.1.41}$$

$$0 \leqslant \tau_i \leqslant 1, \quad i = 0, 1, \ldots, K, \tag{3.1.42}$$

$$0 \leqslant \rho_i \leqslant 1, \quad i = 1, \ldots, K. \tag{3.1.43}$$

The above problem is in general non-convex due to the coupled optimization variables of time allocations and PS ratios and thus cannot be solved optimally. However, we can obtain a near-optimal solution by optimizing $\rho = [\rho_1, \rho_2, \ldots, \rho_K]$ and $\tau = [\tau_0, \tau_1, \ldots, \tau_K]$ iteratively.

Initializing ρ, $P2$ will be equal to finding the optimal value of τ which can be solved following the same steps as in uplink throughput maximization problem in the previous section. In the kth iteration, we first find $\tau^{(k)}$ given $\rho^{(k-1)}$. Having $\tau^{(k)}$, we can then update power splitting ratios by simply taking the derivative of the objective function with respect to ρ. In this case, maximizing the objective function is equivalent to maximizing γ_i^d ($i = 1, \ldots, K$). Hence, we will have

$$\frac{\partial \gamma_i^d}{\partial \rho_i} = \frac{A_i^d \frac{\tau_0^{(k)}}{\tau_i^{(k)}} \left[(C_i^d - B_i^d \frac{\tau_0^{(k)}}{\tau_i^{(k)}}) \rho_i^{(k)2} - 2C_i^d \rho_i^{(k)} + C_i^d \right]}{\left[B_i^d \frac{\tau_0^{(k)}}{\tau_i^{(k)}} \rho_i^{(k)} + C_i^d (1 - \rho_i^{(k)}) \right]^2} = 0, \qquad (3.1.44)$$

thus,

$$(C_i^d - B_i^d \frac{\tau_0^{(k)}}{\tau_i^{(k)}}) \rho_i^{(k)2} - 2C_i^d \rho_i^{(k)} + C_i^d = 0, \qquad (3.1.45)$$

which results in

$$\rho_i^{(k)} = \frac{C_i^d - \sqrt{B_i^d C_i^d \frac{\tau_0^{(k)}}{\tau_i^{(k)}}}}{C_i^d - B_i^d \frac{\tau_0^{(k)}}{\tau_i^{(k)}}}. \qquad (3.1.46)$$

Note that $\rho_i^{(k)} = \frac{C_i^d + \sqrt{B_i^d C_i^d \tau_0^{(k)} / \tau_i^{(k)}}}{C_i^d - B_i^d \tau_0^{(k)} / \tau_i^{(k)}}$ is infeasible because we will have $\rho_i^{(k)} > 1$ for $C_i^d > B_i^d \tau_0^{(k)} / \tau_i^{(k)}$ and $\rho_i^{(k)} < 0$ for $C_i^d < B_i^d \tau_0^{(k)} / \tau_i^{(k)}$ none of which is a feasible value for $\rho_i^{(k)}$.

Repeating the above process, we update τ and ρ in each iteration until both converge to a predefined accuracy.

3.1.4 Numerical Results

In this section, we evaluate the performance of our proposed DH-WPCN through numerical simulations. Channel power gains are modeled as $|h_i|^2 = \theta_{1,i} D_{1,i}^{-\alpha_1}$ and $|g_i|^2 = \theta_{2,i} D_{2,i}^{-\alpha_2}$ with $D_{1,i}$ representing the distance between the HAP and the ith

Fig. 3.3 Throughput vs.
HAP's transmit power

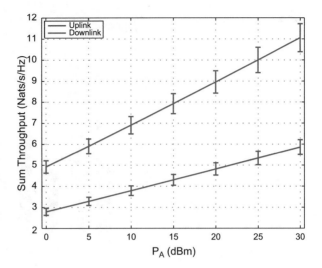

relay and $D_{2,i}$ referring to the distance between the ith relay and the ith user. $\theta_{1,i}$ and $\theta_{2,i}$ indicate the short-term fading which are assumed to be Rayleigh distributed, i.e., $\theta_{1,i}$ and $\theta_{2,i}$ are independent exponential random variables with mean unity. α_1 and α_2 are path-loss exponents. For simplicity, We assume that $\alpha_1 = \alpha_2 = \alpha$, and $\alpha = 2$ unless otherwise specified. Other parameters are set as follows: $\eta_i = 1$, $P_{U,i} = 20dBm$, and $\sigma_H^2 = \sigma_{r,i}^2 = \sigma_{U,i}^2 = -70dBm/Hz$, $\forall i \in \{1, \ldots, K\}$. The results have been averaged over 1000 simulation runs and the 95% confidence intervals are also shown.

Figure 3.3 shows the effect of the HAP's transmit power on uplink and downlink sum-throughput. It is assumed that there are two users in the network with $D_{1,i} = D_{2,i} = 10m$ ($i = 1, 2$). As we can see in Fig. 3.3, both uplink and downlink throughput increase when the HAP's transmit power increases. With higher HAP's transmit power, the relays can harvest more energy and the power of the amplified signal is increased accordingly; therefore, the throughput gets better with increasing HAP's power. Another important observation is that downlink throughput increases faster than uplink throughput. That's because in downlink communication, increasing HAP's transmit power not only results in more harvested energy at the relays but also boosts the power of the HAP's transmitted information signal which also improves SNR and throughput.

In Fig. 3.4, we investigate the effect of the number of users on total uplink and downlink throughput. Here, the transmit power of the HAP is fixed at $P_A = 30dBm$. It is observed that both uplink and downlink sum-throughput are non-decreasing with the number of users in the network.[3] This can be clarified as follows: Suppose

---

[3]Increasing the number of users means that the number of relays is also increased because we assume a dedicated relay for each user.

Fig. 3.4 Throughput vs. number of users

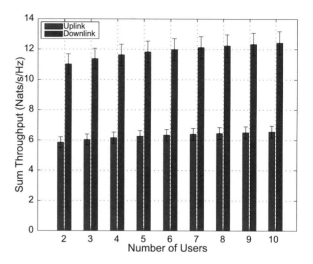

that there are K users in the network and the optimal time allocation and the maximum sum-throughput are denoted as τ^* and $\mathcal{T}^*$, respectively. Now, we add one more user to the network and recalculate the optimal time allocation and sum-throughput. Suppose that τ' and $\mathcal{T}'$ are the new calculated time allocation and sum-throughput, respectively. Now, if $\mathcal{T}' < \mathcal{T}^*$, we can set the time duration of the newly-added user to 0 and allocate time-slots to other users according to τ^*. In this case, the sum-throughput of the DH-WPCN will be equal to $\mathcal{T}^*$. This contradicts our assumption that τ' is the optimal time allocation which results in maximum possible throughput. Hence, $\mathcal{T}' \geqslant \mathcal{T}^*$.

However, Fig. 3.4 shows that the rate at which uplink and downlink total throughput increase becomes lower when we continue adding new users to the network. According to Fig. 3.5, adding new users reduces the optimal time for energy transfer (τ_0) because increasing the number of users implies that more data transmission time is needed which in consequence decreases the energy transfer duration. As energy transfer time gets shorter, users harvest less amount of energy in the first phase which leads to throughput reduction. Therefore, the throughput boost offered by incrementing the number of users is being neutralized by the shortened energy harvesting time of the users. Nevertheless, as explained earlier, both uplink and downlink throughputs are non-decreasing with the number of network users.

What's more, Figs. 3.3 and 3.4 demonstrate that the downlink sum-throughput is greater than the uplink sum-throughput. The main reason behind this observation is that in downlink communication, the effective transmission time of R_i is τ_i while only half of the ith time slot is dedicated to R_i's transmission in uplink communication which makes the effective transmission time of R_i equal to $\tau_i/2$. This less effective transmission time results in lower throughput in uplink communication.

Next, we want to investigate the doubly near-far problem in our proposed model. To this end, we consider a two-user DH-WPCN, in which the users are 10 meters

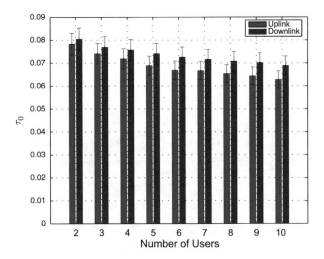

Fig. 3.5 Energy transfer time vs. number of users

away from their corresponding relays (i.e., $D_{2,1} = D_{2,2} = 10m$). We also fix the location of R_1 at 10 meters from the HAP (i.e., $D_{1,1} = 10m$). Now we vary the location of the second relay (R_2) from $D_{1,2} = 10m$ to $D_{1,2} = 20m$.[4] $P_A = 30dBm$ and other simulation parameters are the same as before.

Figure 3.6 shows the throughput ratio $\mathcal{T}_2/\mathcal{T}_1$ as a function of $D_{1,2}/D_{1,1}$ in both uplink and downlink directions. It is observed that increasing the distance between R_2 and the HAP drastically decreases both uplink and downlink throughput ratio, however, we can see that uplink throughput ratio decreases at a faster rate than downlink throughput ratio. This arises from the doubly near-far-problem which appears only in uplink communication. Indeed, in uplink communication, the relay which is located further from the HAP harvests less energy than the other relay, but has to transmit with more power. As a result of this, less time is allocated for information transmission of the user whose relay is far from the HAP leading to a lower throughput for this user. In other words, the throughput of the second user is sacrificed for the sake of maximizing the total throughput.

If we assume that the distance between the HAP and the users is fixed and the relays are flexibly positioned between the HAP and the users, we can mitigate unfairness by adjusting relay locations for a more even allocation of throughput among different users. For example, the relay whose corresponding user is further away from the HAP can be placed nearer to the HAP in order to alleviate the doubly near-far problem. Indeed, such a user still suffers from its long first-hop (user-to-relay) distance; however, as the second-hop (relay-to-HAP) plays a more important

---

[4]Here, changing the location of R_2 does not alter the distance between U_2 and R_2, i.e., $D_{2,2} = 10m$ regardless of the value of $D_{1,2}$ because the objective of this simulation (Fig. 3.6) is merely to observe the doubly near-far problem by examining R_2 in different locations.

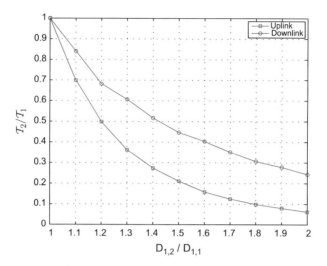

Fig. 3.6 Throughput ratio for a two-user DH-WPCN

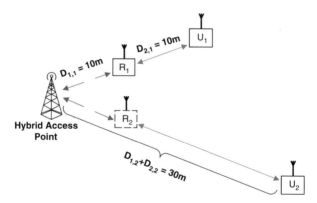

Fig. 3.7 A two-user DH-WPCN

role in individual users' throughputs, it is more desirable to decrease the distance between the relay and the HAP for the users who are more prone to getting unfair throughput share (i.e., further users).

An example is plotted in Fig 3.7, where $D_{1,1} = 10m$, $D_{2,1} = 10m$, and $D_{1,2} + D_{2,2} = 30m$. Other simulation parameters are the same as before. We vary the location of R_2 between U_2 and the HAP from $D_{1,2} = 10m$ to $D_{1,2} = 20m$. Figure 3.8 shows that when R_2 is nearer to the HAP, throughput unfairness between the users is less extreme. In fact, as the doubly near-far problem is related to the second-hop of uplink communication, the location of the relay nodes is very important in determining the fairness level in the network.

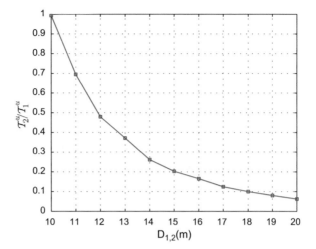

Fig. 3.8 Uplink throughput ratio vs. $D_{1,2}$ when $D_{1,2} + D_{2,2} = 30m$

Next, we want to demonstrate the importance of optimizing time allocations in uplink and downlink communications. For this purpose, we propose suboptimal time allocations and investigate the performance enhancement that our optimal solution yields compared to the proposed suboptimal ones.

The objective of the suboptimal schemes is to allocate equal transmission time to all users. It is easy to show that the objective functions in (3.1.20) and (3.1.40) are monotonically increasing functions of τ_0. This indicates that the time constraints in (3.1.21) and (3.1.41) must be met with equality. Otherwise, we can always increase τ_0 to achieve a higher throughput. Therefore, for the suboptimal solutions we will have

$$\tau_i = \frac{1 - \tau_0}{K} \quad i = 1, \ldots, K. \tag{3.1.47}$$

Replacing (3.1.47) into (3.1.20) and (3.1.40), the uplink throughput maximization problem will turn to a one-variable optimization problem which can be easily solved. For the downlink throughput maximization problem, τ_0 and $\rho = [\rho_1, \rho_2, \ldots, \rho_K]$ are optimization variables. Similar to what we did in the previous section, we iteratively update τ_0 and ρ until satisfactory convergence is obtained.

Figures 3.9 and 3.10 show uplink and downlink sumthroughput, respectively, as a function of the HAP's transmit power for optimal and suboptimal time allocations. In this simulation, there are two users in the DH-WPCN with $D_{1,1} = D_{2,1} = D_{2,2} = 10m$ and $D_{1,2} = 20m$. $\alpha = 3$ and other simulation parameters are the same as in previous simulations.

We can see that using optimal time allocations always results in higher uplink and downlink throughputs. The figures also illustrate that the throughput gap between optimal and suboptimal solutions increases with increasing P_A. In other words, the

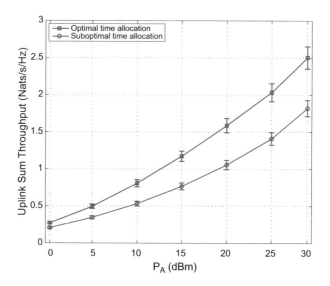

Fig. 3.9 Uplink throughput performance with optimal and suboptimal time allocations

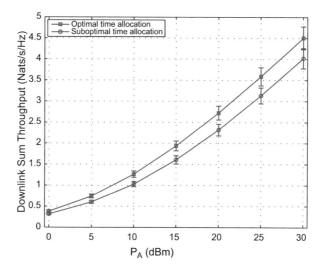

Fig. 3.10 Downlink throughput performance with optimal and suboptimal time allocations

importance of using optimized time allocations becomes more perceptible in larger values of the HAP's transmit power.

Finally, we investigate the performance gain achieved by using optimal power splitting ratios in downlink communication. Again, $\alpha = 3$ and there are two users in the network with $D_{1,1} = D_{2,1} = D_{2,2} = 10m$ and $D_{1,2} = 20m$. Figure 3.11 depicts downlink throughput performance with and without using optimal power splitting ratios. We can see that fixed power splitting ratios (i.e., $\rho_i = \rho$, $\forall i \in$

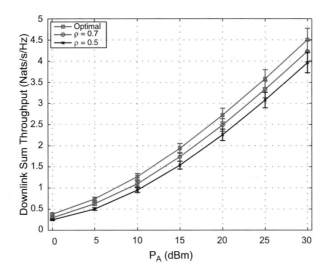

Fig. 3.11 Downlink throughput performance with optimal and fixed power splitting ratios

$\{1, \ldots, K\})$ always offer a lower throughput than the optimized ones. This clarifies that beside optimizing time durations, using optimal ratios for power splitting is also of paramount importance for downlink throughput enhancement.

3.2 Fairness Enhancement in DH-WPCN

In Sect. 3.1, we studied a dual-hop wireless powered communication network (DH-WPCN) and investigated throughput maximization in both uplink and downlink directions. The simulations conducted at the end of Sect. 3.1 revealed a severe fairness problem in terms of individual users' throughput in uplink communication. The user whose relay is located far from the HAP suffers from the so-called doubly near-far problem and is assigned a small transmission time which leads to a significantly low throughput for this user. We proposed to flexibly position the relays between the users and the HAP so as to control the doubly near-far problem in the network. However, it is not always possible to change the location of the relays. Hence, a more robust solution is needed to ensure throughput fairness among network users. The aim of this chapter is to tackle the doubly near-far problem in our proposed DH-WPCN by optimizing time allocations for energy and information transfer. We consider the problem of maximizing the minimum throughput among users in order to guarantee throughput fairness for the users and yet maximize the total throughput. Our findings show that there exists a trade-off between sum-throughput and fairness. While the sum-throughput maximization scheme proposed in the previous chapter demonstrates a better performance in terms of the total uplink throughput, it experiences a low fairness level and the fairness deteriorates with increasing the number of users. On the other hand, the minimum throughput maximization (MTM) approach presented in this chapter preserves fairness regardless of the number of users. The content of this chapter has appeared in [70].

3.2.1   Minimum Throughput Maximization

Similar to the previous section, the system model consists of a HAP, K users, and K energy-constrained relays (Fig. 3.1). HAP and users are assumed to have fixed energy supplies, while the relays rely on harvesting energy from downlink energy signals broadcast by the HAP. As discussed in Sect. 3.1, the achievable uplink throughput from U_i (*i*th user) to the HAP is expressed as

$$T_i = \frac{\tau_i}{2} \log(1 + \gamma_i) \simeq \frac{\tau_i}{2} \log(1 + \frac{A_i \frac{\tau_0}{\tau_i}}{B_i \frac{\tau_0}{\tau_i} + C_i}),\,^5 \tag{3.2.1}$$

where

$$A_i = 2\eta_i P_A P_{U,i} |h_i|^4 |g_i|^2, \tag{3.2.2}$$

$$B_i = 2\eta_i P_A |h_i|^4 \sigma_{r,i}^2, \tag{3.2.3}$$

$$C_i = P_{U,i} |g_i|^2 \sigma_H^2. \tag{3.2.4}$$

We want to guarantee a minimum throughput for each user, and yet maximize their sum-throughput. According to (3.2.1), minimum throughput maximization problem is formulated as follows:

$$(P3): \max \; R$$

$$s.t. \; \frac{\tau_i}{2} log(1 + \frac{A_i \frac{\tau_0}{\tau_i}}{B_i \frac{\tau_0}{\tau_i} + C_i}) \geqslant R \tag{3.2.5}$$

$$\sum_{i=0}^{K} \tau_i \leqslant 1, \tag{3.2.6}$$

$$0 \leqslant \tau_i \leqslant 1, \; i = 0, 1, \ldots, K, \; R \geqslant 0,$$

---

[5]We omit superscript *u* because only uplink communication is investigated here and we do not study downlink communication in this chapter.

where the constraint in (3.2.5) is to ensure a minimum throughput for all users and (3.2.6) is the time constraint, meaning that the time portions for downlink energy transfer and uplink information transmission must be no greater than the length of the transmission block.

Lemma 3.2 *For the optimal solution of the above minimum throughput maximization problem, the constraints in (3.2.5) and (3.2.6) must be met with equality.*

Proof Suppose that $\tau_x = [\tau_{0x}, \tau_{1x}, \ldots, \tau_{Kx}]$ is the optimal solution of $P3$ such that $\sum_{i=0}^{K} \tau_{ix} < 1$. Now, we can form another solution with $\tau_y = [\tau_{0x} + (1 - \sum_{i=0}^{K} \tau_{ix}), \tau_{1x}, \ldots, \tau_{Kx}]$. Since $\mathcal{T}_i(\tau) = \dfrac{\tau_i}{2} log(1 + \dfrac{A_i \tau_0/\tau_i}{B_i \tau_0/\tau_i + C_i})$ is a monotonically increasing function of τ_0 for $\tau_0 \geqslant 0$, $\mathcal{T}_i(\tau_y) > \mathcal{T}_i(\tau_x)$. This contradicts our assumption that τ_x is the optimal solution. Therefore, the optimal solution of $P3$ satisfies (3.2.6) with equality. Similarly, as $\mathcal{T}_i$ is a monotonically increasing function of τ_i for $\tau_i > 0$, maximum R is achieved when all users get equal throughput. Suppose that $U = [U_1, \ldots, U_K]$ is the set of network users and U_n has the minimum throughput among all users, i.e., $\mathcal{T}_i > \mathcal{T}_n$, $\forall i \in \{1, \ldots, K\} \neq n$. Now we can decrease τ_i's and increase τ_n to have a higher minimum throughput. This demonstrates that the optimal solution is obtained when equality holds for the constraint in (3.2.5). ∎

From Lemma 3.2, we know that the optimal solution of $P3$ must satisfy the following equation:

$$R^* = \frac{\tau_i^*}{2} log(1 + \frac{A_i \dfrac{\tau_0^*}{\tau_i^*}}{B_i \dfrac{\tau_0^*}{\tau_i^*} + C_i}), \quad \forall i \in \{1, \ldots, K\}. \tag{3.2.7}$$

For a given R and τ_0, τ_i's are the solution of

$$g(\tau_i, R) = \tau_0, \tag{3.2.8}$$

where

$$g(\tau_i, R) = \frac{C_i \tau_i (e^{2R/\tau_i} - 1)}{A_i - (e^{2R/\tau_i} - 1) B_i}. \tag{3.2.9}$$

Given R, $P3$ is feasible if we can find $\tau = [\tau_0, \tau_1, \ldots, \tau_K]$ such that $\sum_{i=0}^{K} \tau_i \leqslant 1$. In the following, we illustrate the feasibility of this problem with an example (Fig. 3.12).

Consider a DH-WPCN consisting of two users as shown in Fig. 3.12 with $D_{1,1} = 10m$, $D_{1,2} = 20m$ and $D_{2,1} = D_{2,2} = 10m$, where as explained in the previous chapter, $D_{1,i}$ represents the distance between the HAP and the ith relay and $D_{2,i}$ refers to the distance between the ith relay and the ith user. Rayleigh fading

Fig. 3.12 A two-user
DH-WPCN

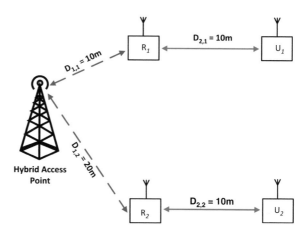

channels are modeled as $|h_i|^2 = \theta_{1,i} D_{1,i}^{-\alpha_1}$ and $|g_i|^2 = \theta_{2,i} D_{2,i}^{-\alpha_2}$, where $\theta_{1,i}$ and $\theta_{2,i}$ are independent exponential random variables with mean unity. Throughout this chapter, $\alpha_1 = \alpha_2 = 2$, unless otherwise stated. Other parameters are as follows: $P_A = 20\,dBm$, $\eta_i = 1$, $P_{U,i} = 20\,dBm$, and $\sigma_H^2 = \sigma_{r,i}^2 = \sigma_{U,i}^2 = -70\,dBm/Hz$, $\forall i \in \{1, \ldots, K\}$.

Figure 3.13 shows $\sum_{i=0}^{2} \tau_i$ versus τ_0 for different values of R. It can be observed that $R = 2.5\ Nats/s/Hz$ is not feasible for the network setup described above because its corresponding $\sum_{i=0}^{2} \tau_i$ graph never meets the line $\sum_{i=0}^{2} \tau_i = 1$. On the other hand, $R = 2\ Nats/s/Hz$ and $R = 1.5\ Nats/s/Hz$ are feasible solutions as we can find $\tau = [\tau_0, \tau_1, \tau_2]$ such that the total time duration is less than or equal to 1. Figure 4.2 also illustrates that the total time duration is shifted upward as we increase R. Using bisection method, R^* is achieved when the corresponding $\sum_{i=0}^{K} \tau_i$ graph is tangent to the line $\sum_{i=0}^{K} \tau_i = 1$ and the point of intersection determines the optimal value of τ_0 for which τ_i's can be found using (3.2.8).

Algorithm 1 summarizes the process of finding τ^* and R^* (Fig. 3.13).

Figure 3.14 plots time allocation ratio (τ_2/τ_1) versus α_1 for the two-user DH-WPCN of Fig. 3.12. Throughout this chapter, "STM" refers to the sum-throughput maximization scheme presented in Chap. 3, while "MTM" represents the minimum throughput maximization approach proposed in this chapter. We can see that the time allocation ratio decreases in STM and increases in MTM with increasing α_1. As α_1 increases, the channels between the HAP and the relays start to degrade, however, R_2 experiences a more severe degradation due to its longer distance to the HAP. STM allocates more time to the user whose relay has better channel conditions in order to maximize the total uplink throughput. This unfairness in time allocation intensifies with increasing α_1 and results in throughput unfairness. According to Fig. 3.15, the throughput ratio between the two users $(\mathcal{T}_2/\mathcal{T}_1)$ drastically decreases in STM when α_1 is increased and we can see that U_2 gets negligible throughput when $\alpha = 3$. MTM keeps users' throughputs at the same level by assigning U_2 a greater transmission time (Figs. 3.14, 3.15, 3.16, 3.17, 3.18).

Algorithm 1 Minimum throughput maximization

Initialize $R_{min} = 0$, R_{max}.
repeat
$\quad R = 0.5(R_{min} + R_{max})$.
$\quad \tau_0 = 0$.
$\quad$**while** $\tau_0 \in [0, 1]$ **do**
$\quad\quad$ Compute τ_i's using (3.2.8).
$\quad\quad$ Compute $\sum_{i=0}^{K} \tau_i$.
$\quad\quad$**if** $0 \leqslant 1 - \sum_{i=0}^{K} \tau_i < \epsilon$ **then**
$\quad\quad\quad \tau_0^* = \tau_0$.
$\quad\quad\quad \tau_i^* = \tau_i$.
$\quad\quad$**end if**
$\quad\quad \tau_0 \leftarrow \tau_0 + \Delta$.
$\quad$**end while**
$\quad$**if** no $\tau = [\tau_0, \tau_1, \dots, \tau_K]$ was found such that $\sum_{i=0}^{K} \tau_i \leqslant 1$ **then**
$\quad\quad R_{max} \leftarrow R$.
$\quad$**else**
$\quad\quad R_{min} \leftarrow R$.
$\quad$**end if**
until $R_{max} - R_{min} < \delta$.
$R^* = R$.

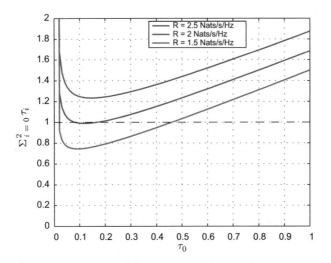

Fig. 3.13 Total time duration versus τ_0 for the network setup of Fig. 4.1

3.2.2   Numerical Results

In this section, we present simulation results to evaluate the performance of our minimum throughput maximization algorithm in terms of throughput and fairness. To this end, we compare the proposed scheme with the STM approach discussed in Chap. 3 which aimed to maximize the sum-throughput of DH-WPCN. Channel

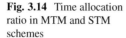

Fig. 3.14 Time allocation ratio in MTM and STM schemes

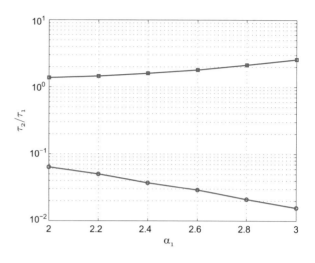

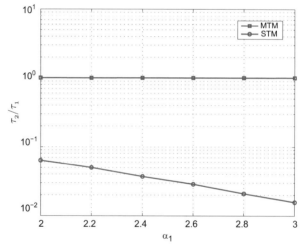

Fig. 3.15 Throughput ratio in MTM and STM schemes

models and simulation parameters are the same as those used in Sect. 3.2.1 unless otherwise specified. Simulation results have been averaged over 1000 runs and the 95% confidence intervals are also shown.

Figure 3.16 shows the throughput performance of STM and MTM schemes versus HAP's transmit power when we have two users in the network with the setup depicted in Fig. 3.12. It can be observed that the normalized throughput of STM is higher than R^* which is the normalized throughput of the MTM scheme. That's because the STM method sacrifices the second user's throughput for the sake of sum-throughput maximization. As we can see in the figure, the main part of the maximum sum-throughput belongs to the first user and the second user gets only a

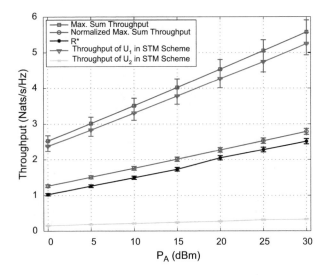

Fig. 3.16 Throughput vs. HAP's transmit power in STM and MTM schemes

Fig. 3.17 Effect of the number of users on throughput

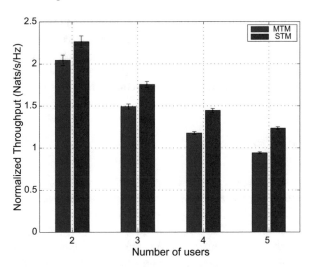

small share of the total throughput. MTM provides all users with equal throughput in order to realize throughput fairness.

Figures 3.17 and 3.18 plot normalized throughput and fairness index versus number of users (K), where $D_{1,i} = \dfrac{20m}{K} \times i$ and $D_{2,i} = 10m$, $\forall i \in \{1, \ldots, K\}$. $P_A = 20dBm$ and other simulation parameters are the same as before. Fairness index is calculated as $F = \dfrac{(\sum_{i=1}^{K} x_i)^2}{K \sum_{i=1}^{K} x_i^2}$ [71], where x_i is the throughput of the ith user. Figure 3.17 shows that in both schemes, normalized throughput decreases

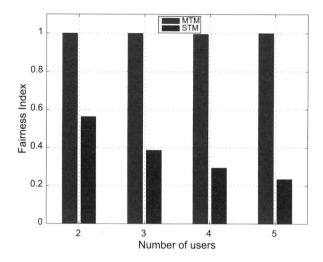

Fig. 3.18 Effect of the number of users on fairness index

with increasing K and STM outperforms MTM in terms of throughput. However, the lower throughput of the MTM approach is made up by its higher fairness level as illustrated in Fig. 3.18. According to Fig. 3.18, STM suffers from unfair throughput allocation and this gets worse when K is increased. On the other hand, MTM always maintains fairness as all users achieve the same throughput.

3.3 Summary

In this chapter, we investigated fairness enhancement in a dual-hop WPCN (DH-WPCN), where energy-constrained relays assist the communication between energy-stable users and the HAP using an amplify-and-forward strategy. To overcome the doubly near-far problem which occurs due to unequal distance of the relays from the HAP, we studied the minimum throughput maximization problem and optimized time allocations to achieve throughput fairness among users. According to the numerical results, there is a trade-off between throughput and fairness in the DH-WPCN and the improved fairness is achieved at the expense of a reduction in throughput. The proposed fairness-enhancing algorithm retains the highest fairness level with increasing the number of users and in different channel conditions which makes it suitable for applications that require equal throughput allocation for all users.

References

1. R. Minerva, A. Biru, and D. Rotondi, "Towards a Definition of the Internet of Things (IoT)," May 2015, [online] Available: http://iot.ieee.org/definition.html.
2. H. Sundmaeker, P. Guillemin, P. Friess, and S. Woelffle, "Vision and Challenges for Realising the Internet of Things," Cluster of European Research Projects on the Internet of Things, March 2010.
3. A. McEwen and H. Cassimally, "Designing the Internet of Things," John Wiley and Sons, 2014.
4. A. Al-Fuqaha, M. Guizani, M. Mohammadi, M. Aledhari, and M. Ayyash, "Internet of Things: A Survey on Enabling Technologies, Protocols, and Applications," *IEEE Communications Surveys and Tutorials*, vol. 13, no. 4, pp. 2347–2376, Fourth Quarter 2015.
5. L. Tan and N. Wang, "Future Internet: The Internet of Things," *3rd International Conference on Advanced Computer Theory and Engineering(ICACTE)*, August 2010, pp. V5-376 - V5-380.
6. W. Ye, J. Heidemann, and D. Estrin, "An Energy-Efficient MAC Protocol for Wireless Sensor Networks," *Proceedings of the Twenty-First Annual Joint Conference of the IEEE Computer and Communications Societies*, June 2002, pp. 1567–1576.
7. T. Van Dam and K. Langendoen, "An Adaptive Energy-Efficient MAC Protocol for Wireless Sensor Networks," *Proceedings of the 1st International Conference on Embedded Networked Sensor Systems*, November 2003, pp. 171–180.
8. A. El-Hoiydi and J-D. Decotignie, "WiseMAC: An Ultra Low Power MAC Protocol for Multi-hop Wireless Sensor Networks," *Proceedings of the First International Workshop, ALGOSENSORS*, July 2004, pp. 18–31.
9. J. Polastre, J. Hill, and D. Culler, "Versatile Low Power Media Access for Wireless Sensor Networks," *Proceedings of the 2nd International Conference on Embedded Networked Sensor Systems*, November 2004, pp. 95–107.
10. Y. Yu, R. Govindan, and D. Estrin, "Geographical and Energy Aware Routing: a recursive data dissemination protocol for wireless sensor networks," *Technical Report UCLA/CSD-TR-01-0023*, UCLA Computer Science Department, May 2001.
11. D. Ganesan, R. Govindan, S. Shenker, and D. Estrin, "Highly-Resilient, Energy-Efficient Multipath Routing in Wireless Sensor Networks," *ACM SIGMOBILE Mobile Computing and Communications Review*, vol. 5, no. 4, pp. 11–25, October 2001.
12. V. Raghunathan, A. Kansal, J. Hsu, J. Friedman, and M. Srivastava, "Design Considerations for Solar Energy Harvesting Wireless Embedded Systems," *Fourth International Symposium on Information Processing in Sensor Networks (IPSN)*, April 2005, pp. 457–462.
13. Y. K. Tan and S. K. Panda, "Optimized Wind Energy Harvesting System Using Resistance Emulator and Active Rectifier for Wireless Sensor Nodes," *IEEE Transactions on Power Electronics*, vol. 26, no. 1, pp. 38–50, January 2011.
14. M. A. Weimer, T. S. Paing, and R. A. Zane, "Remote area wind energy harvesting for low-power autonomous sensors," *37th IEEE Power Electronics Specialists Conference*, 2006, pp. 1–5.
15. H. Kulah and K. Najafi, "Energy Scavenging From Low-Frequency Vibrations by Using Frequency Up-Conversion for Wireless Sensor Applications," *IEEE Sensors Journal*, vol. 8, no. 3, pp. 261–268, March 2008.
16. X. Lu and S-H. Yang, "Thermal Energy Harvesting for WSNs," *IEEE International Conference on Systems, Man and Cybernetics*, October 2010, pp. 3045–3052.
17. X. Lu, P. Wang, D. Niyato, D. I. Kim, and Z. Han, "Wireless Networks With RF Energy Harvesting: A Contemporary Survey," *IEEE Communications Surveys and Tutorials*, vol. 17, no. 2, pp. 757–789, Second Quarter 2015.
18. P. Kamalinejad, C. Mahapatra, Z. Sheng, S. Mirabbasi, V. C. M. Leung, and Y. L. Guan, "Wireless Energy Harvesting for the Internet of Things," *IEEE Communications Magazine*, vol. 53, no. 6, pp. 102–108, June 2015.

19. T. Soyata, L. Copeland, and W. Heinzelman, "RF Energy Harvesting for Embedded Systems: A Survey of Tradeoffs and Methodology," *IEEE Circuits and Systems Magazine*, vol. 16, no. 1, pp. 22–57, First Quarter 2016.
20. S. Bi, C. K. Ho, and R. Zhang, "Wireless Powered Communication: Opportunities and Challenges," *IEEE Communications Magazine*, vol. 53, no. 4, pp. 117–125, April 2015.
21. R. Zhang, R. G. Maunder, and L. Hanzo, "Wireless Information and Power Transfer: From Scientific Hypothesis to Engineering Practice," *IEEE Communications Magazine*, vol. 53, no. 8, pp. 99–105, August 2015.
22. L. R. Varshney, "Transporting Information and Energy Simultaneously," *IEEE International Symposium on Information Theory (ISIT)*, July 2008, pp. 1612–1616.
23. P. Grover and A. Sahai, "Shannon meets Tesla: Wireless information and power transfer," *IEEE International Symposium on Information Theory (ISIT)*, June 2010, pp. 2363–2367.
24. A. M. Fouladgar and O. Simeone, "On the Transfer of Information and Energy in Multi-User Systems," *IEEE Communications Letters*, vol. 16, no. 11, pp. 1733–1736, November 2012.
25. Z. Xiang and M. Tao, "Robust Beamforming for Wireless Information and Power Transmission," *IEEE Wireless Communications Letters*, vol. 1, no. 4, pp. 372–375, August 2012.
26. R. Zhang and C. K. Ho, "MIMO Broadcasting for Simultaneous Wireless Information and Power Transfer," *IEEE Transactions on Wireless Communications*, vol. 12, no. 5, pp. 1989–2001, May 2013.
27. A. A. Nasir, X. Zhou, S. Durrani, and R. A. Kennedy, "Relaying Protocols for Wireless Energy Harvesting and Information Processing," *IEEE Transactions on Wireless Communications*, vol. 12, no. 7, pp. 3622–3636, July 2013.
28. S. Bi, Y. Zeng, and R. Zhang, "Wireless Powered Communication Networks: An Overview," *IEEE Wireless Communications*, vol. 23, no. 2, pp. 10–18, April 2016.
29. H. Ju and R. Zhang, "Throughput Maximization in Wireless Powered Communication Networks," *IEEE Transactions on Wireless Communications*, vol. 13, no. 1, pp. 418–428, January 2014.
30. H. Ju and R. Zhang, "Optimal Resource Allocation in Full-Duplex Wireless-Powered Communication Network," *IEEE Transactions on Communications*, vol. 62, no. 10, pp. 3528–3540, October 2014.
31. X. Kang, C. K. Ho, and S. Sun, "Full-Duplex Wireless-Powered Communication Network With Energy Causality," *IEEE Transactions on Wireless Communications*, vol. 14, no. 10, pp. 5539–5551, October 2015.
32. L. Liu, R. Zhang, and K-C. Chua, "Multi-Antenna Wireless Powered Communication With Energy Beamforming," *IEEE Transactions on Communications*, vol. 62, no. 12, pp. 4349–4361, December 2014.
33. G. Yang, C. K. Ho, R. Zhang, and Y. L. Guan, "Throughput Optimization for Massive MIMO Systems Powered by Wireless Energy Transfer," *IEEE Journal on Selected Areas in Communications*, vol. 33, no. 8, pp. 1640–1650, August 2015.
34. D. Hwang, D. I. Kim, and T-J. Lee, "Throughput Maximization for Multiuser MIMO Wireless Powered Communication Networks," *IEEE Transactions on Vehicular Technology*, vol. 65, no. 7, pp. 5743–5748, July 2016.
35. S. Lee and R. Zhang, "Cognitive Wireless Powered Network: Spectrum Sharing Models and Throughput Maximization," *IEEE Transactions on Cognitive Communications and Networking*, vol. 1, no. 3, pp. 335–346, September 2015.
36. S. S. Kalamkar, J. P. Jeyaraj, A. Banerjee, and K. Rajawat, "Resource Allocation and Fairness in Wireless Powered Cooperative Cognitive Radio Networks," *IEEE Transactions on Communications*, vol. 64, no. 8, pp. 3246–3261, August 2016.
37. Y. L. Che, L. Duan, and R. Zhang, "Spatial Throughput Maximization of Wireless Powered Communication Networks," *IEEE Journal on Selected Areas in Communications*, vol. 33, no. 8, pp. 1534–1548, August 2015.
38. H. Ju and R. Zhang, "User Cooperation in Wireless Powered Communication Networks," *IEEE Global Communications Conference*, December 2014, pp. 1430–1435.

39. H. Chen, Y. Li, J. L. Rebelatto, B. F. Uchoa-Filho, and B. Vucetic, "Harvest-Then-Cooperate: Wireless-Powered Cooperative Communications," *IEEE Transactions on Signal Processing*, vol. 63, no. 7, pp. 1700–1711, April 2015.
40. N. Tesla, "Method of Regulating Apparatus For Producing Currents of High Frequency," *U.S. Patent No. 568,178*, September 1896.
41. N. Tesla, "Apparatus for Transmitting Electrical Energy,"*U.S. Patent No. 1,119,732*, December 1914.
42. W. C. Brown, " Experiments Involving a Microwave Beam to Power and Position a Helicopter," *IEEE Transactions on Aerospace and Electronic Systems*, vol. AES-5, no. 5, pp. 692–702, September 1969.
43. W. C. Brown, " The History of Power Transmission by Radio Waves," *IEEE Transactions on Microwave Theory and Techniques*, vol. 32, no. 9, pp. 1230–1242, September 1984.
44. C-S. Wang, G. A. Covic, and O. H. Stielau, "Power Transfer Capability and Bifurcation Phenomena of Loosely Coupled Inductive Power Transfer Systems," *IEEE Transactions on Industrial Electronics*, vol. 51, no. 1, pp. 148–157, February 2004.
45. B. L. Cannon, J. F. Hoburg, D. D. Stancil, and S. C. Goldstein "Magnetic Resonant Coupling As a Potential Means for Wireless Power Transfer to Multiple Small Receivers," *IEEE Transactions on Power Electronics*, vol. 24, no. 7, pp. 1819–1825, July 2009.
46. X. Lu, P. Wang, D. Niyato, D. I. Kim, and Z. Han, "Wireless Charging Technologies: Fundamentals, Standards, and Network Applications," *IEEE Communications Surveys and Tutorials*, vol. 18, no. 2, pp. 1413–1452, Second Quarter 2016.
47. S. D. Barman, A. W. Reza, N. Kumar, M. E. Karim, A. B. Munir, "Wireless powering by magnetic resonant coupling: Recent trends in wireless power transfer system and its applications," *Renewable and Sustainable Energy Reviews*, vol. 51, pp. 1525–1552, November 2015.
48. X. Lu, P. Wang, D. Niyato, and Z. Han, "Resource Allocation in Wireless Networks with RF Energy Harvesting and Transfer," *IEEE Network*, vol. 29, no. 6, pp. 68–75, November-December 2015.
49. H. T. Friis, "A Note on a Simple Transmission Formula," *Proceedings of the IRE*, vol. 34, no. 5, pp. 254–256, May 1946.
50. D. Mishra, S. De, S. Jana, S. Basagni, K. Chowdhury, and W. Heinzelman, "Smart RF Energy Harvesting Communications: Challenges and Opportunities," *IEEE Communications Magazine*, vol. 53, no. 4, pp. 70–78, April 2015.
51. M. Pinuela, P. D. Mitcheson, and S. Lucyszyn, "Ambient RF Energy Harvesting in Urban and Semi-Urban Environments," *IEEE Transactions on Microwave Theory and Techniques*, vol. 61, no. 7, pp. 2715–2726, July 2013.
52. A. Ghazanfari, H. Tabassum, and E. Hossain, "Ambient RF Energy Harvesting in Ultra-Dense Small Cell Networks: Performance and Trade-offs," *IEEE Wireless Communications*, vol. 23, no. 2, pp. 38–45, April 2016.
53. S. Lee, R. Zhang, and K. Huang, "Opportunistic Wireless Energy Harvesting in Cognitive Radio Networks," *IEEE Transactions on Wireless Communications*, vol. 12, no. 9, pp. 4788–4799, September 2013.
54. D. T. Hoang, D. Niyato, P. Wang, and D. I. Kim, "Opportunistic Channel Access and RF Energy Harvesting in Cognitive Radio Networks," *IEEE Journal on Selected Areas in Communications*, vol. 32, no. 11, pp. 2039–2052, November 2014.
55. Powercast, [Online] Available: www.powercastco.com
56. I. Krikidis, S. Timotheou, S. Nikolaou, G. Zheng, D. W. K. Ng, and R. Schober, "Simultaneous Wireless Information and Power Transfer in Modern Communication Systems," *IEEE Communications Magazine*, vol. 52, no. 11, pp. 104–110, November 2014.
57. X. Zhou, R. Zhang, and C. K. Ho, "Wireless Information and Power Transfer: Architecture Design and Rate-Energy Tradeoff," *IEEE Transactions on Communications*, vol. 61, no. 11, pp. 4754–4767, November 2013.

58. H. Lee, C. Song, S-H. Choi, and I. Lee, "Outage Probability Analysis and Power Splitter Designs for SWIPT Relaying Systems with Direct Link," *IEEE Communications Letters*, 2016.
59. K. Huang and E. Larsson, "Simultaneous Information and Power Transfer for Broadband Wireless Systems," *IEEE Transactions on Signal Processing*, vol. 61, no. 23, pp. 5972–5986, December 2013.
60. I. Krikidis, "Simultaneous Information and Energy Transfer in Large-Scale Networks with/without Relaying," *IEEE Transactions on Communications*, vol. 62, no. 3, pp. 900–912, March 2014.
61. I. Krikidis, S. Sasaki, S. Timotheou, and Z. Ding, "A Low Complexity Antenna Switching for Joint Wireless Information and Energy Transfer in MIMO Relay Channels," *IEEE Transactions on Communications*, vol. 62, no. 5, pp. 1577–1587, May 2014.
62. C-F. Liu, M. Maso, S. Lakshminarayana, C-H. Lee, T. Q. S. Quek, "Simultaneous Wireless Information and Power Transfer Under Different CSI Acquisition Schemes," *IEEE Transactions on Wireless Communications*, vol. 14, no. 4, pp. 1911–1926, April 2015.
63. L. Mohjazi, I. Ahmed, S. Muhaidat, M. Dianati, and M. Al-Qutayri, "Downlink Beamforming for SWIPT Multi-User MISO Underlay Cognitive Radio Networks," *IEEE Communications Letters*, 2016.
64. A. Sabharwal, P. Schniter, D. Guo, D. W. Bliss, S. Rangarajan, and R. Wichman, "In-Band Full-Duplex Wireless: Challenges and Opportunities," *IEEE Journal on Selected Areas in Communications*, vol. 32, no. 9, pp. 1637–1652, September 2014.
65. H. Ju, K. Chang, and M-S. Lee, "In-Band Full-Duplex Wireless Powered Communication Networks," *17th International Conference on Advanced Communication Technology (ICACT)*, December 2014, pp. 23–27.
66. Q. Wu, M. Tao, D. W. K. Ng, W. Chen, and R. Schober, "Energy-Efficient Resource Allocation for Wireless Powered Communication Networks," *IEEE Transactions on Wireless Communications*, vol. 15, no. 3, pp. 2312–2327, March 2016.
67. P. Ramezani and A. Jamalipour, "Throughput Maximization in Dual-Hop Wireless Powered Communication Networks," Accepted to appear in *IEEE Transactions on Vehicular Technology*, 2017.
68. K. T. Phan, T. Le-Ngoc, S. A. Vorobyov, and C. Tellambura, "Power Allocation in Wireless Multi-User Relay Networks," *IEEE Transactions on Wireless Communications*, vol. 8, no. 5, pp. 2535–2545, May 2009.
69. S. Boyd and L. Vanderberghe, "Convex Optimization," *Cambridge University Press*, 2004.
70. P. Ramezani and A. Jamalipour, "Fairness Enhancement in Dual-Hop Wireless Powered Communication Networks," Accepted in *IEEE International Conference on Communications (IEEE ICC2017)*, May 2017.
71. R. K. Jain, D-M. W. Chiu, and W. R. Hawe, "A Quantitative Measure of Fairness and Discrimination For Resource Allocation in Shared Computer Systems," *Technical Report TR-301*, DEC Research Report, 1984.
72. C. Song, K-J. Lee, and I. Lee, "MMSE-Based MIMO Cooperative Relaying Systems: Closed-Form Designs and Outage Behavior," *IEEE Journal on Selected Areas in Communications*, vol. 30, no. 8, pp. 1390–1401, September 2012.
73. H-B. Kong, C. Song, H. Park, and I. Lee, "A New Beamforming Design for MIMO AF Relaying Systems With Direct Link," *IEEE Transactions on Communications*, vol. 62, no. 7, pp. 2286–2295, July 2014.
74. N. Lee and R. W. Heath Jr, "Advanced Interference Management Technique: Potentials and Limitations," *IEEE Wireless Communications*, vol. 23, no. 3, pp. 30–38, June 2016.

Chapter 4
Future Direction of Wireless Powered Communications

Abstract This final chapter draws general conclusions and points out several areas for future research. The specific contributions of this work summarized at the end of each chapter are not repeated here.

4.1 Final Remarks

The steady growth in wireless communications, fostered by the prosperity of mobile services, has resulted in an unprecedented awareness of the potential of PLS to significantly strengthen the security level of current systems. The fundamental idea of PLS is to exploit the inherent randomness of fading and interference of wireless channels to restrain the amount of information that can be gleaned at the bit level by a passive eavesdropper. Although having been studied extensively in various kinds of networks such as wireless sensor networks, cellular networks, cognitive networks, it is fair to acknowledge that the practicality of PLS in WPCNs is still an open problem.

Chapter 2 has been primarily concerned with investigating challenges of PLS raised especially in WPCNs, with an emphasis on the design and evaluation of practical transmission schemes exploiting the coordination of information transmission and wireless power transfer. In particular, we developed and analyzed two practical secure transmission protocols for helper-assisted wiretap channels based on cooperative jamming. Both proposed protocols exploit the controllability of RF energy harvesting to allow the wireless-powered node to transmit with optimal transmit power. In addition, the secrecy performance of these protocols relies heavily on the availability of CSI; the specifically designed jamming signal takes advantage of the estimated CSI, either perfect or imperfect, to judiciously direct interference towards the eavesdropper. Our analysis of the protocols showed that, in the case with perfect CSI and multiple eavesdroppers, a less capable helper with only few antennas can maintain confidentiality; however, in another case when the CSI estimation is imperfect, the occurrence of secrecy outage increases notably as the estimation error grows, even under the existence of only one single eavesdropper.

© Springer Nature Switzerland AG 2019

A. Jamalipour, Y. Bi, *Wireless Powered Communication Networks*,

https://doi.org/10.1007/978-3-319-98174-1_4

Nevertheless, the proposed protocol under imperfect CSI still provides superior performance over the existing solution, owing to the enabled full-duplex capability of the helper.

In Chap. 3, we proposed a DH-WPCN with one HAP and a number of relays and users. Assuming that the users have fixed energy supplies and the relays need to harvest energy from RF transmission of the HAP, we presented uplink and downlink communication protocols. Optimal values of parameters for maximizing the total throughput of the network in both directions were found. Specifically, we formulated uplink and downlink sum-throughput maximization problems to find optimal time allocation in both uplink and downlink communications as well as optimal power splitting factors in downlink communication. The convex structure of the uplink throughput maximization problem allowed us to obtain the optimal value of the time-slot durations for energy and information transfer in closed form, while in downlink throughput maximization, we used iterations for finding a near-optimal solution due to the non-convexity of the problem. We evaluated the uplink and downlink throughput performance of our proposed schemes via simulations and identified the existence of the doubly near-far problem in uplink communication which results in extremely unfair throughput distribution among the users. Due to the dependence of each user's achievable throughput on its corresponding relay's distance from the HAP, we proposed to dynamically adjust the location of the relays to attain a more balanced throughput allocation in the network. Numerical results confirmed that the position of the relays has a major impact on users' throughputs and the severity of the throughput unfairness can be controlled by changing relays' placements. Furthermore, we investigated a more robust solution for tackling the aforementioned fairness issue and developed a fairness enhancement scheme to provide all users with equal throughput. We formulated a minimum throughput maximization problem and proposed a novel algorithm for finding the maximum common throughput of the users plus the optimal time allocation for achieving the maximum level of fairness. We also conducted simulations to compare the performance of the proposed fairness-improving scheme with the strategy presented in Chap. 3. Simulation results revealed a throughput-fairness trade-off in our DH-WPCN implying that the fairness is achieved at the cost of total throughput reduction. Therefore, depending on the network requirements in terms of sum-throughput and fairness, either the strategy proposed in Section or the scheme presented in Chap. 4 can be chosen as the optimal policy.

4.2 The Way Forward

4.2.1 Future Research Directions

The scope for future work on PLS that can be extended from this work is extensive. A few selected directions are discussed below.

- In this work, only the single-jammer case was considered. When multiple jammers present, coordinated or uncoordinated jamming schemes need to be explored. In a coordinated jamming scheme, multiple jammers may generate and transmit jamming signals in a coordinated manner to avoid interfering with the intended receiver. Since the jammers are usually spatially distributed, such coordination requires that the global CSI of each jammer should be collected, and the collaborative jamming weights should be optimized. Evidently, this leads to extra communication overhead. Alternatively, uncoordinated jamming schemes are of great potential to overcome the limitation of coordinated jamming schemes, albeit with the potential of yielding undesired interference at the intended receiver.
- Studies on secure communications for large-scale WPCNs need to be carried out. Unlike point-to-point wiretap channels considered in this work, the communication between nodes in large-scale WPCNs depends strongly on the location and the interactions between nodes. Particularly, stochastic geometry tools such as a Poisson point process model may be adopted to describe the spatial location of legitimate nodes and/or eavesdroppers.
- Very few studies on PLS are concerned with the joint design of cryptographic techniques. Indeed, a deeper understanding of the interplay between PLS and classic cryptographic security is an important, rich, but unexplored area for further investigation.

In addition, we outline some future directions for enhancing the DH-WPCN presented in Chap. 3.

- In Chap. 3, we proposed adjustment of the relays' locations in order to alleviate the doubly near-far problem in uplink communication. As discussed earlier, for the users who are placed further away from the HAP, it is more desirable to locate their corresponding relays closer to the HAP so that the allocated throughput for these users would be comparable to the ones nearer to the HAP. Nevertheless, it is obvious that a relay cannot be very distant from its corresponding user because it must be able to successfully receive the information signal of the user. In light of this, optimizing the location of the relay nodes can be an interesting problem for future research. For instance, the optimal location of the relays can be determined for maximizing the throughput fairness under the first- and the second-hop outage constraints.
- In the uplink communication of the proposed DH-WPCN, the first half of each transmission time-slot is dedicated to one of the users to send information to its relay. During this time, the HAP and the other $K - 1$ relays and users remain idle. As an enhancement to our proposed uplink communication protocol, the HAP can continue transferring wireless energy in the beginning half of all transmission slots so that the relays which are not receiving information from their corresponding users can add to their harvested energy.
- In our DH-WPCN, we assumed that there is no direct link between the users and the HAP. The proposed schemes can be extended by considering the presence

of a link between each user and the HAP to achieve diversity and multiplexing gains [1, 2].

• The proposed DH-WPCN can be enhanced to FD scenarios, where the HAP is able to simultaneously transfer energy and receive information on the same frequency band. This strategy can boost the harvestable energy of the relays which consequently improves the throughput. Also, activating FD operation at the relays will enable them to harvest energy from the HAP's transmitted signals while they are receiving information from their users.

• The proposed model assumes that the HAP is equipped with one single antenna. By applying a multi-antenna HAP and taking advantage of the beamforming technique, the energy transfer efficiency can be improved. In this case, doubly near-far problem can also be mitigated because the beamforming vector plays an important role in the amount of available energy at each relay. As such, designing beamforming weights in the favor of further relays can offset their harsh channel conditions and bring a more balanced throughput distribution.

• The DH-WPCN presented can be extended to more complicated network setups. For instance, it would be interesting to consider a large-scale network similar to the one proposed in [3], where multiple HAPs and a large number of users and relays are present. In this setup, cooperative energy beamforming can be used to improve the energy transfer efficiency at the target relays by jointly optimizing the energy signal waveforms of the HAPs, thus forming a virtual antenna array. Integrating DH-WPCN with cognitive radio network can be another possible direction for future research.

• This research extended the baseline one-hop energy and information transfer model to a scenario in which information transmission is carried out in two hops. This model can be further extended to include two-hop energy transfer as well. Assuming that both the relays and the users are energy-constrained, dual-hop energy transfer may be utilized such that the relays transfer a portion of their harvested energy to their corresponding users. Studying the resource allocation problem in this new DH-WPCN setup can be considered for future works.

• In reality, the number of relays may be different from the number of users. This can be taken into consideration in future research. Relay selection mechanisms can be implemented to determine which relay should be in charge of data forwarding for each user. Also, if more than one user is assigned to a relay, it is important to specify how much time and energy must be allocated for information transmission/reception of each of them.

4.2.2 A Closer Look into WPCN from the IoT Point of View

IoT consists of heterogeneous devices with different constraints, requirements, and capabilities which makes it impossible to apply one strategy to all scenarios. For example, the required energy for the operation of a low-power sensor node is not comparable to the amount needed for powering a mobile phone. Moreover, different IoT devices may be in different conditions in terms of the volume of data to be transmitted, delay constraints, QoS requirements, etc. One device may experience long delay (e.g., several weeks) between two transmission cycles while the other has to transmit once every second. Also, some devices may have critical data (e.g., urgent medical information) which need to be prioritized for timely transmission. These dissimilarities between different IoT components have to be taken into account in designing resource allocation schemes.

Further, information exchange between network devices should also be considered in IoT. Allocation of time and energy resources in this case may be more complicated than the scenario in which devices only communicate with the HAP. The communication between devices also provides them with an extra energy harvesting source. The receiver can use the transmitter's signal for the dual purpose of information decoding and energy harvesting, while other devices treat this signal as an ambient RF energy source; even the transmitter itself can collect energy from its own RF signal when operating in the full-duplex mode. In a WPCN having multiple transmitter-receiver pairs, interference management methods [4] will be helpful for minimizing the effect of interference at the receivers.

Multi-hop communication is also necessary in future scalable IoT networks, where a number of nodes cooperate with each other to deliver data to the designated node (e.g., the HAP). Although our dual-hop WPCN protocol can provide insights for extending the baseline WPCN to multi-hop networks, extensive research has still to be done in this area for developing multi-hop WPCNs which perfectly fit into IoT. In such a multi-hop WPCN setup, in addition to the information routing in conventional multi-hop networks, energy routing can also be exploited to transfer energy in multiple hops. In this regard, multi-path energy routing is useful for increasing RF energy harvesting efficiency and extending the network size [5]. Coexistence of energy-constrained and energy-stable devices is also imaginable in IoT in which case energy-stable ones can participate in energy transfer process to help feeding energy-limited devices, thus, mitigating the need for deploying additional energy sources and saving on the related costs.

To sum up, seamless integration of WPCN into IoT calls for more intelligent solutions. These solutions can be a combination of the strategies reviewed and proposed in this work as well as new techniques which cater to the needs of future IoT networks.

References

1. C. Song, K.-J. Lee, and I. Lee, "MMSE-Based MIMO Cooperative Relaying Systems: Closed-Form Designs and Outage Behavior," IEEE J. Sel. Areas Commun., vol. 30, no. 8, pp. 1390–1401, Sep. 2012.
2. H.-B. Kong, C. Song, H. Park, and I. Lee, "A New Beamforming Design for MIMO AF Relaying Systems With Direct Link," IEEE Trans. Commun., vol. 62, no. 7, pp. 2286–2295, Jul. 2014.
3. Y. L. Che, L. Duan, and R. Zhang, "Spatial Throughput Maximization of Wireless Powered Communication Networks," IEEE J. Sel. Areas Commun., vol. 33, no. 8, pp. 1534–1548, Aug. 2015.

4. N. Lee and R. W. Heath Jr, "Advanced Interference Management Technique: Potentials and Limitations," IEEE Wireless Commun., vol. 23, no. 3, pp. 30–38, Jun. 2016.
5. D. Mishra, S. De, S. Jana, S. Basagni, K. Chowdhury, and W. Heinzelman, "Smart RF Energy Harvesting Communications: Challenges and Opportunities," IEEE Commun. Mag., vol. 53, no. 4, pp. 70–78, Apr. 2015.

Appendix A
Proofs for Chap. 2

A.1 Proof of Proposition 2.2

According to the total probability theorem, the secrecy outage probability p_{so} can be expressed as

$$p_{so} = \Pr\left\{\max\{\gamma_{e_n}\} > \beta | \zeta = \zeta_{PT}\right\} \Pr\{\zeta = \zeta_{PT}\}$$
$$+ \Pr\left\{\max\{\gamma_{e_n}\} > \beta | \zeta = \zeta_{IT}\right\} \Pr\{\zeta = \zeta_{IT}\}. \qquad (A.1.1)$$

We first evaluate the secrecy outage in PT mode. Recall that no secret information is transmitted in PT mode, $\gamma_{e,n} = 0$, $\forall n$. Since $R_t > R_s$, $\beta = 2^{R_t - R_s} - 1$ is positive. It hence can be inferred that $\Pr\left\{\max\{\gamma_{e,n}\} > \beta | \zeta = \zeta_{PT}\right\} = 0$. Therefore, we have $\Pr\left\{\max\{\gamma_{e_n}\} > \beta | \zeta = \zeta_{PT}\right\} \Pr\{\zeta = \zeta_{PT}\} = 0$.

Next, we evaluate the secrecy outage in IT mode.

$$\Pr\left\{\max\{\gamma_{e_n}\} > \beta | \zeta = \zeta_{IT}\right\} \Pr\{\zeta = \zeta_{IT}\}$$
$$= \Pr\{\zeta = \zeta_{IT}\} - \Pr\left\{\max\{\gamma_{e_n}\} \le \beta | \zeta = \zeta_{IT}\right\} \Pr\{\zeta = \zeta_{IT}\}$$
$$= p_{tx} - \Pr\left\{(\max\{\gamma_{e_n}\} \le \beta) \bigcap (\zeta = \zeta_{IT})\right\}$$
$$= p_{tx} - \sum_{i=1}^{L} \pi_i \Pr\left\{(H_{ab} \ge U) \bigcap \left(H_{ab} \ge \frac{\alpha\sigma_b^2 T L}{iC}\right)\right\}$$
$$= p_{tx} - \sum_{i=1}^{L} \pi_i \left(\underbrace{\Pr\left\{\left(H_{ab} \ge \frac{\alpha\sigma_b^2 T L}{iC}\right) \bigcap \left(U < \frac{\alpha\sigma_b^2 T L}{iC}\right)\right\}}_{\ell_{2a}}\right)$$

A. Jamalipour, Y. Bi, *Wireless Powered Communication Networks*,
https://doi.org/10.1007/978-3-319-98174-1

$$+ \Pr \left\{ (H_{ab} \geq U) \bigcap \underbrace{\left(U \geq \frac{\alpha \sigma_b^2 T L}{i C} \right) \right\} \right)}_{\ell_{2b}}. \tag{A.1.2}$$

where

$$U := \max_{n \in \Phi_E} \left\{ \frac{\frac{\alpha \sigma_b^2}{\beta} |h_{ae_n}|^2}{\frac{P_0}{K-1} ||\mathbf{g}_{pe_n}^\dagger \mathbf{W}||^2 + \sigma_e^2} \right\}. \tag{A.1.3}$$

To proceed, we need to derive the CDF of the random variable U which depends on $|h_{ae_n}|^2$ and $||\mathbf{g}_{pe_n}^\dagger \mathbf{W}||^2$. For notation simplicity, we define $X_{e_n} := \alpha \sigma_b^2 |h_{ae_n}|^2/\beta$, $Y_{e_n} := P_0 ||\mathbf{g}_{pe_n}^\dagger \mathbf{W}||^2/(K-1)$, and $Z_{e_n} := X_{e_n}/(Y_{e_n} + \sigma_e^2)$. Recall that all wireless links are assumed to be Rayleigh fading, therefore, X_{e_n} follows an exponential distribution with the PDF given by

$$f_{X_{e_n}}(x) = \frac{\beta}{\alpha \sigma_b^2 \lambda_{ae}} \exp\left(-\frac{\beta}{\alpha \sigma_b^2 \lambda_{ae}} x \right). \tag{A.1.4}$$

On the other hand, $||\mathbf{g}_{pe_n}||^2$ is a sum of i.i.d. exponentially distributed random variables. Since $\mathbf{W}$ is a unitary matrix, $||\mathbf{g}_{pe_n}^\dagger \mathbf{W}||^2$ remains a sum of i.i.d. exponentially distributed random variables. Therefore, Y_{e_n} follows a Gamma distribution with parameters $(K-1, P_0 \lambda_{pe}/(K-1))$. The PDF of Y_{e_n} is given by

$$f_{Y_{e_n}}(y) = \frac{y^{K-2} \exp\left(-\frac{K-1}{P_0 \lambda_{pe}} y \right)}{\Gamma(K-1) \left(\frac{P_0 \lambda_{pe}}{K-1} \right)^{K-1}}. \tag{A.1.5}$$

Therefore, based on Lemma 2.1, the CDF of Z_{e_n} can be obtained as

$$\begin{aligned}
F_{Z_{e,n}}(z) &= \Pr\left(\frac{X_{e_n}}{Y_{e_n} + \sigma_e^2} < z \right) \\
&= \int_0^\infty \int_0^{zy+z\sigma_e^2} f_{X_{e_n}}(x) f_{Y_{e_n}}(y) \mathrm{d}x \mathrm{d}y \\
&= 1 - \exp\left(-\frac{\sigma_e^2 \beta}{\alpha \sigma_b^2 \lambda_{ae}} z \right) \left(\frac{K-1}{\frac{P_0 \lambda_{pe} \beta}{\alpha \sigma_b^2 \lambda_{ae}} z + K - 1} \right)^{K-1} \\
&= 1 - e^{-\phi z} \left(\frac{K-1}{\psi z + K - 1} \right)^{K-1}. \tag{A.1.6}
\end{aligned}$$

where for notation simplicity ϕ and ψ are defined respectively as

$$\phi := \frac{\sigma_e^2 \beta}{\alpha \sigma_b^2 \lambda_{ae}} \quad \text{and} \quad \psi := \frac{P_0 \lambda_{pe} \beta}{\alpha \sigma_b^2 \lambda_{ae}}. \tag{A.1.7}$$

Consequently, the CDF of U can be obtained as

$$F_U(u) = \Pr\left\{ \max\{Z_{e_n}\} < u \right\} = \left(F_{Z_{e_n}}(u) \right)^N. \tag{A.1.8}$$

Now we are ready to resolve ℓ_{2a} and ℓ_{2b}.

$$\ell_{2a} = \int_{\frac{\alpha \sigma_b^2 TL}{iC}}^{\infty} \int_0^{\frac{\alpha \sigma_b^2 TL}{iC}} f_U(U) f_{H_{ab}}(H_{ab}) \, dU \, dH_{ab} = \exp\left(-\frac{\alpha \sigma_b^2 TL}{i \lambda_{ab} C} \right) F_U\left(\frac{\alpha \sigma_b^2 TL}{iC} \right).$$
$$\tag{A.1.9}$$

and

$$\ell_{2b} = \int_{\frac{\alpha \sigma_b^2 TL}{iC}}^{\infty} \int_{\frac{\alpha \sigma_b^2 TL}{iC}}^{H_{ab}} f_U(U) f_{H_{ab}}(H_{ab}) \, dU \, dH_{ab}$$

$$= \int_{\frac{\alpha \sigma_b^2 TL}{iC}}^{\infty} \left[F_U(H_{ab}) - F_U\left(\frac{\alpha \sigma_b^2 TL}{iC} \right) \right] f_{H_{ab}}(H_{ab}) \, dH_{ab}$$

$$= \int_{\frac{\alpha \sigma_b^2 TL}{iC}}^{\infty} F_U(H_{ab}) f_{H_{ab}}(H_{ab}) \, dH_{ab} - F_U\left(\frac{\alpha \sigma_b^2 TL}{iC} \right) \exp\left(-\frac{\alpha \sigma_b^2 TL}{i \lambda_{ab} C} \right).$$
$$\tag{A.1.10}$$

The resultant integral term in (A.1.10) can be calculated as follows,

$$\int_{\frac{\alpha \sigma_b^2 TL}{iC}}^{\infty} F_U(H_{ab}) f_{H_{ab}}(H_{ab}) \, dH_{ab}$$

$$= \int_{\frac{\alpha \sigma_b^2 TL}{iC}}^{\infty} \left[1 - e^{-\phi H_{ab}} \left(\frac{K-1}{\psi H_{ab} + K - 1} \right)^{K-1} \right]^N \frac{1}{\lambda_{ab}} \exp\left(-\frac{H_{ab}}{\lambda_{ab}} \right) \, dH_{ab}$$

$$= \int_{\frac{\alpha\sigma_b^2 TL}{iC}}^{\infty} \left[1 + \sum_{m=1}^{N} \binom{N}{m} \left(-e^{-\phi H_{ab}} \left(\frac{K-1}{\psi H_{ab}+K-1} \right)^{K-1} \right)^m \right] \frac{1}{\lambda_{ab}} \exp\left(-\frac{H_{ab}}{\lambda_{ab}} \right) dH_{ab}$$

$$= \exp\left(-\frac{\alpha\sigma_b^2 TL}{i\lambda_{ab}C} \right) + \sum_{m=1}^{N} \binom{N}{m} \frac{(-1)^m}{\lambda_{ab}} \left(\frac{K-1}{\psi} \right)^{m(K-1)}$$

$$\times \int_{\frac{\alpha\sigma_b^2 TL}{iC}}^{\infty} \frac{\exp\left(-\left(\phi m + \frac{1}{\lambda_{ab}} \right) H_{ab} \right)}{\left(H_{ab} + \frac{K-1}{\psi} \right)^{m(K-1)}} dH_{ab}$$

$$= \exp\left(-\frac{\alpha\sigma_b^2 TL}{i\lambda_{ab}C} \right) + \sum_{m=1}^{N} \binom{N}{m} \frac{(-1)^m}{\lambda_{ab}} \left(\frac{K-1}{\psi} \right)^{m(K-1)}$$

$$\times \exp\left(-\left(\phi m + \frac{1}{\lambda_{ab}} \right) \frac{\alpha\sigma_b^2 TL}{iC} \right) \Psi\left(m(K-1), \phi m + \frac{1}{\lambda_{ab}}, \frac{\alpha\sigma_b^2 TL}{iC} + \frac{K-1}{\psi} \right).$$

$$(A.1.11)$$

where the last step is obtained from [38, Eq. (3.353.2)]. Substituting (A.1.11) into (A.1.10), and combining with (A.1.9), (A.1.2), (2.1.25) and (A.1.1), after some basic mathematical manipulation, we obtain the final closed-form expression of the secrecy outage probability in (2.1.29).

A.2   Proof of Proposition 2.3

From the definition of secrecy outage probability given in (2.2.78), by applying the total probability theorem, P_{so}^{AnJ} can be expressed as

$$P_{so}^{AnJ} = \underbrace{\Pr\{C_s < R_s | \Phi = \Phi_d\} \Pr\{\Phi = \Phi_d\}}_{\ell_1} + \underbrace{\Pr\{C_s < R_s | \Phi = \Phi_o\} \Pr\{\Phi = \Phi_o\}}_{\ell_2}$$

$$(A.2.1)$$

First, we evaluate the secrecy outage in DEH mode. Recall that no secret data is transmitted in DEH blocks, γ_D and γ_E hence both equal zero, and further, C_s equals 0. As R_s is positive, it can be inferred that $\Pr\{C_s < R_s | \Phi = \Phi_d\} = 1$. Therefore we have $\ell_1 = \Pr\{\Phi = \Phi_d\}$. Invoking the independence between the channel condition and the energy condition, and also combining (2.2.53) and (2.2.76), we can obtain

$$\ell_1 = 1 - \Pr\{(C_{SD} \geq R_s) \cap (\varepsilon[k] \geq E_{th})\}$$

$$= 1 - \left(1 - F_{H_{SD}}\left(\frac{2^{R_s} - 1}{P_S/\sigma_D^2}\right)\right)\sum_{i=\tau}^{L}\xi_{FD,i} \qquad (A.2.2)$$

Next, we evaluate the secrecy outage in OEH mode. Considering that C_s and C_{SD} are not independent with each other, but both are independent with the energy random variables, we recast ℓ_2 as,

$$\ell_2 = \Pr\{(C_s < R_s) \cap (C_{SD} \geq R_s) \cap (\varepsilon[k] \geq E_{th})\}$$

$$= \underbrace{\Pr\{(C_s < R_s) \cap (C_{SD} \geq R_s)\}}_{\ell_A}\sum_{i=\tau}^{L}\xi_{FD,i} \qquad (A.2.3)$$

Substituting (2.2.33), (2.2.78) and (2.2.80) into (A.2.3), and performing basic mathematical manipulations, we obtain

$$\ell_A = \Pr\left\{\left(H_{SD} < \frac{(1 + \gamma_E)2^{R_s} - 1}{\kappa_1}\right) \cap \left(H_{SD} \geq \frac{2^{R_s} - 1}{\kappa_2}\right)\right\}$$

$$= \Pr\left\{\frac{2^{R_s} - 1}{\kappa_2} \leq H_{SD} < \frac{(1 + \gamma_E)2^{R_s} - 1}{\kappa_1}\right\}$$

$$= \int_0^\infty \int_{\frac{2^{R_s} - 1}{\kappa_2}}^{\frac{(1+\gamma_E)2^{R_s} - 1}{\kappa_1}} f_{H_{SD}}(H_{SD})f_{\gamma_E}(\gamma_E)\mathrm{d}H_{SD}\,\mathrm{d}\gamma_E$$

$$= \int_0^\infty F_{H_{SD}}\left(\frac{(1 + \gamma_E)2^{R_s} - 1}{\kappa_1}\right)f_{\gamma_E}(\gamma_E)\,\mathrm{d}\gamma_E - F_{H_{SD}}\left(\frac{2^{R_s} - 1}{\kappa_2}\right) \qquad (A.2.4)$$

where $f_{H_{SD}}(\cdot)$ represents the PDF of H_{SD} and

$$\kappa_1 := \frac{P_S}{(1 - \rho)P_J\sigma_{err}^2/(N_t - 1) + \sigma_D^2}. \qquad (A.2.5)$$

$$\kappa_2 := \frac{P_S}{\sigma_D^2}. \qquad (A.2.6)$$

Substituting (2.2.52) together with (2.2.85) into (A.2.4), and applying [38, Eq. (3.352.4) and Eq. (3.353.2)] to solve the resultant integrals, we have

$$\int\limits_0^\infty F_{H_{SD}}\left(\frac{(1+\gamma_E)2^{R_s}-1}{\kappa_1}\right) f_{\gamma_E}(\gamma_E)\,\mathrm{d}\gamma_E$$

$$= 1 - \int_0^\infty \exp\left(-\frac{(1+\gamma_E)2^{R_s}-1}{\kappa_1\Omega_{SD}}\right) f_{\gamma_E}(\gamma_E)\,\mathrm{d}\gamma_E$$

$$=1 - \frac{\sigma_E^2\, e^{-\frac{2^{R_s}-1}{\kappa_1\Omega_{SD}}}}{P_S\Omega_{SE}} \int_0^\infty \exp\left(-\left(\frac{2^{R_s}}{\kappa_1\Omega_{SD}}+\frac{\sigma_E^2}{P_S\Omega_{SE}}\right)\gamma_E\right)\left(\frac{N_t-1}{\varphi\gamma_E+N_t-1}\right)^{N_t-1}\,\mathrm{d}\gamma_E$$

$$-\varphi e^{-\frac{2^{R_s}-1}{\kappa_1\Omega_{SD}}} \int_0^\infty \exp\left(-\left(\frac{2^{R_s}}{\kappa_1\Omega_{SD}}+\frac{\sigma_E^2}{P_S\Omega_{SE}}\right)\gamma_E\right)\left(\frac{N_t-1}{\varphi\gamma_E+N_t-1}\right)^{N_t}\,\mathrm{d}\gamma_E$$

$$=\begin{cases}1-\left(\frac{\sigma_E^2\beta_1}{P_S\Omega_{SE}}\Psi_1(1,\mu,\frac{N_t-1}{\varphi})-\varphi\beta_1^2\Psi_1(2,\mu,\frac{N_t-1}{\varphi})\right)\exp\left(-\frac{2^{R_s}-1}{\kappa_1\Omega_{SD}}\right) & \text{if } N_t=2\\[2mm] 1-\left(\frac{\sigma_E^2\beta_1^{N_t-1}}{P_S\Omega_{SE}}\Psi_2(N_t-1,\mu,\beta_1)-\varphi\beta_1^{N_t}\Psi_2(N_t,\mu,\beta_1)\right)\exp\left(-\frac{2^{R_s}-1}{\kappa_1\Omega_{SD}}\right) & \text{if } N_t>2\end{cases}$$

$$(A.2.7)$$

Substituting (A.2.7) into (A.2.4), we can obtain

$$\ell_A = \begin{cases}\exp\left(-\frac{2^{R_s}-1}{\kappa_2\Omega_{SD}}\right) & \text{if } N_t=2\\[1mm] \quad-\frac{\sigma_E^2\beta_1}{P_S\Omega_{SE}}\exp\left(-\frac{2^{R_s}-1}{\kappa_1\Omega_{SD}}\right)\Psi_1(1,\mu,\frac{N_t-1}{\varphi})\\[1mm] \quad-\varphi\beta_1^2\exp\left(-\frac{2^{R_s}-1}{\kappa_1\Omega_{SD}}\right)\Psi_1(2,\mu,\frac{N_t-1}{\varphi})\\[1mm] \exp\left(-\frac{2^{R_s}-1}{\kappa_2\Omega_{SD}}\right) & \text{if } N_t\geq 3\\[1mm] \quad-\frac{\sigma_E^2\beta_1^{N_t-1}}{P_S\Omega_{SE}}\exp\left(-\frac{2^{R_s}-1}{\kappa_1\Omega_{SD}}\right)\Psi_2(N_t-1,\mu,\beta_1)\\[1mm] \quad-\varphi\beta_1^{N_t}\exp\left(-\frac{2^{R_s}-1}{\kappa_1\Omega_{SD}}\right)\Psi_2(N_t,\mu,\beta_1)\end{cases}$$

$$(A.2.8)$$

Therefore, replacing ℓ_A in (A.2.3) with (A.2.8) and combining with (A.2.2), after some basic mathematical manipulation, we obtain the final result in (2.2.86), thus completing the proof.

A.3 Proof of Corollary 2.2

From the definition of the probability of non-zero secrecy capacity given in (2.2.79), by applying the total probability theorem, P_{nzsc}^{AnJ} can be expressed as

$$P_{nzsc}^{\text{AnJ}} = \underbrace{\Pr\{C_s>0|\Phi=\Phi_d\}\Pr\{\Phi=\Phi_d\}}_{\ell_3} + \underbrace{\Pr\{C_s>0|\Phi=\Phi_o\}\Pr\{\Phi=\Phi_o\}}_{\ell_4}$$

$$(A.3.1)$$

Again, as no secret is transmitted in DEH mode, $\Pr\{C_s > 0 | \Phi = \Phi_d\} = 0$. Therefore, we have $\ell_3 = 0$. And similar to (A.2.3), we can recast ℓ_4 as

$$\ell_4 = \Pr\{(C_s > 0) \cap (C_{SD} \geq R_s) \cap (\varepsilon[k] \geq E_{th})\}$$

$$= \underbrace{\Pr\{(C_s > 0) \cap (C_{SD} \geq R_s)\}}_{\ell_B} \sum_{i=\tau}^{L} \xi_{FD,i} \tag{A.3.2}$$

Substituting (2.2.33), (2.2.79) and (2.2.80) into (A.3.2), and performing basic mathematical manipulations, we obtain

$$\ell_B = \Pr\left\{\left(H_{SD} > \frac{\gamma_E}{\kappa_1}\right) \cap \left(H_{SD} \geq \frac{2^{R_s} - 1}{\kappa_2}\right)\right\}$$

$$= \Pr\left\{\left(H_{SD} > \frac{\gamma_E}{\kappa_1}\right) \cap \left(\frac{\gamma_E}{\kappa_1} \geq \frac{2^{R_s} - 1}{\kappa_2}\right)\right\}$$

$$+ \Pr\left\{\left(H_{SD} \geq \frac{2^{R_s} - 1}{\kappa_2}\right) \cap \left(\frac{\gamma_E}{\kappa_1} < \frac{2^{R_s} - 1}{\kappa_2}\right)\right\}$$

$$= \int_{\frac{\kappa_1(2^{R_s}-1)}{\kappa_2}}^{\infty} \int_{\frac{\gamma_1}{\kappa_1}}^{\infty} f_{H_{SD}}(H_{SD}) f_{\gamma_E}(\gamma_E) \mathrm{d}H_{SD} \, \mathrm{d}\gamma_E$$

$$+ \int_{0}^{\frac{\kappa_1(2^{R_s}-1)}{\kappa_2}} \int_{\frac{(2^{R_s}-1)}{\kappa_2}}^{\infty} f_{H_{SD}}(H_{SD}) f_{\gamma_E}(\gamma_E) \mathrm{d}H_{SD} \, \mathrm{d}\gamma_E \tag{A.3.3}$$

Substituting (2.2.52) together with (2.2.84) and (2.2.85) into (A.3.4), and applying [38, Eq. (3.352.4) and (3.353.2)] to solve the resultant integrals, we derive ℓ_B as

$$\ell_B = \begin{cases} \left(\frac{\sigma_E^2}{P_S \Omega_{SE}} \Psi_1(1, \mu_2, \beta_1 + \beta_2) + \Psi_1(2, \mu_2, \beta_1 + \beta_2)\right) & \text{if } N_t = 2 \\ \times \varphi^{-1} e^{-\beta_2 \mu_2} + \exp\left(-\frac{2^{R_s}-1}{\kappa_2 \Omega_{SD}}\right) F_{\gamma_E}(\beta_2) \\ \\ \left(\frac{\sigma_E^2}{P_S \Omega_{SE}} \Psi_2(N_t - 1, \mu_2, \beta_1 + \beta_2) + (N_t - 1)\Psi_2(N_t, \mu_2, \beta_1 + \beta_2)\right) & \text{if } N_t \geq 3 \\ \times \beta_1^{N_t - 1} e^{-\beta_2 \mu_2} + \exp\left(-\frac{2^{R_s}-1}{\kappa_2 \Omega_{SD}}\right) F_{\gamma_E}(\beta_2) \end{cases}$$

$$\tag{A.3.4}$$

Therefore, substituting ℓ_B into (A.3.2), after some basic mathematical manipulation, we obtain the final result in (2.2.89), thus completing the proof.

Printed in the United States
By Bookmasters